vanessa newsome

electric fashion

drawing fashion flats with ADOBE® ILLUSTRATOR®

CS5

electric fashion
drawing fashion flats with ADOBE* ILLUSTRATOR*
CS5

Published by Electric Fashion Media, Inc.
Address: 645 West 9th Street, Unit 110-285
Los Angeles, California 90015-1640, USA
Website: www.electricfashionmedia.com
Email: info@electricfashionmedia.com
Facebook: electric fashion media, inc.

First edition 2009
Second edition 2010

Cover Design: Stuart Patterson at Colorola

Publisher's Cataloging-in-Publication Data

Newsome, Vanessa

Electric Fashion
Drawing Fashion Flats With Adobe Illustrator
by Vanessa Newsome
ISBN 978-1-4507-3238-3

Trademarks
Electric Fashion is a registered trademark of
Electric Fashion Media, Incorporated.
Adobe, Illustrator and Photoshop are registered
trademarks of Adobe Systems, Incorporated in
the United States and/or other countries.
SnapFashun is a registered trademark of
SnapFashun, Incorporated.

acknowledgments

many thanks

To my fashion friends and colleagues who've inspired my creative and professional growth over the years. Kristen Gloviak, Delia Hodson, Patrick Wood, Wendy Bendoni, Marva Martindale, Drew Bernstein (Lippy), Chaz Austin, and of course Billy Glazier and Normand Duplessies.

A big thank you to the fashion design department and its fantastic team of instructors at the Fashion Institute of Design & Merchandising (FIDM) and to all my students, past, present and future, thank you too.

Finally, a special acknowledgment to the most influential person in my life hands down, thank you A. Watcher for believing that I could do this. This book is dedicated to my one and only DEJ.

Vanessa Newsome

about the author

Vanessa Newsome is a fashion department, lead instructor at the Los Angeles based Fashion Institute of Design & Merchandising (FIDM). She studied fashion design at FIDM, San Francisco and holds a Bachelor of Science degree in visual communications. Her talents in the fashion industry encompasses 17 years of experience in the areas of fashion design, fashion illustration, trend forecasting and graphic design. Ms. Newsome's fashion flats and illustrations have been published in several trend forecasting publications and design software programs. She has also co-authored several books on the subject of computer aided fashion design and looks forward to authoring many more.

preface

welcome

Once upon a time I sat down at a computer with the goal of mastering the skill of drawing fashion electronically. It was a challenge, I must admit, but once I was up and going I was excited to see the end results of my efforts. Now, 17 years later *electric fashion* is here for the student as well as the professional who has an interest in mastering the skill of drawing fashion on computer and enjoying every minute of it.

how electric fashion works

Chapters 1, 2 and 3 are the core chapters. It is in these chapters that you will get a clear understanding of Adobe Illustrator and how it is used to draw fashion flats.

Chapter 4 is a setup chapter. It is a chapter that demonstrates how to setup an Adobe Illustrator workspace before you begin drawing. Each time you set out to draw a particular fashion flat you will want to first setup your workspace as shown in chapter 4.

Chapters 5 is the first fashion flat, lesson chapter and each chapter thereafter is a lesson on drawing a particular fashion flat or detail. The lesson chapters are arranged in skill order from beginner, with detailed instructions, to interme- diate, with less detailed instructions and in later lesson chapters (**6,7, 8...**) the instructions are more advanced.

The techniques demonstrated in *electric fashion* have worked for me for many years but are not assumed to be the only way to draw fashion flats. Electric Fashion is a guide to drawing fashion flats and is intended to inspire your own style and creativity.

quick tips

Quick tips, found throughout the pages of *electric fashion*, offer efficient, time saving techniques for a particular Adobe Illustrator command or function. When it comes to using Adobe Illustrator you'll find that for every one way there is to do something, for example zoom the page, there are likely five other ways to do it. Quick tips point out the best options.

mac and windows

The lessons in *electric fashion* were designed on a PC computer using Windows operating system. Not to worry Mac users, Adobe Illustra- tor from Mac to Windows varies only slightly. The difference lies in basic operation (Exit vs. Quit) and keyboard shortcuts. See page 15 for a list of commonly used Mac and Windows keyboard shortcuts.

keyboard shortcuts

Keyboard shortcuts are important to drawing fashion flats with efficiency and accuracy. You will find that the lessons in *electric fashion* rely greatly on keyboard shortcuts and so it is recommended that you get familiar with your computer's keyboard.

Adobe CS5

The lessons in *electric fashion* were designed in **Adobe Illustrator CS5**. Those of you with earlier versions of Illustrator (CS2,CS3 or CS4) should note that some Electric Fashion screen shots will vary slightly. CS5 users should save files as earlier versions of Illustrator (CS3 or CS4) in order to open and edit the documents in earlier versions. Refer to your software's documenta- tion for more information.

electric fashion CD-rom

The *electric fashion* CD-rom, located on the back page, contains Adobe Illustrator* documents that are used throughout the lessons in this book. The CD-rom is both Mac and Windows compatible.

The **Croquis** folder contains the women's croqui which is used as a guide for creating various fashion flats.

In the **Draw Practice** folder you'll find practice files that can be used to enhance your electronic drawing skills.

Also on the CD-rom, in the **Fabrication** folder, are fabric and pattern swatches. The swatches can be used to fabricate your fashion flats.

You may require a simple letter size page, a tabloid page or several pages whith an Adobe Illustrator document. You will find pages of all sizes and setup in the **Page Layout** folder.

* Adobe Illustrator software is not included with the accompanying CD-rom. Visit www.adobe.com to purchase the Adobe software.

enjoy

Learning this unique skill should be enjoyable most of all. Remember to keep your shoulders relaxed and take a break from the screen every 20 minutes or so. You are a fashion designer first and a computer wiz second and so it is good practice to plan your designs with traditional paper and pencil (quick sketches) and then draw the fashion flat using the demonstrated techniques.

contents

contents in detail

contents in detail (continued)

chapter one

a good view

When your checking out the view in Adobe Illustrator know that it is a drawing program that comes fully stocked with drawing, editing and painting tools of all sorts. You may think of it as an electronic version of an artist's work area that comes with paper, pencils, pens, paintbrushes and more. The fact that Adobe Illustrator is electronic (computerized) is what makes the drawing of a fashion flat and exploring its design a simple task.

Becoming familiar with the Adobe Illustrator workspace, tools, panels, menus, and keyboard shortcuts is the key to mastering its many features. Once you've learned your way around you'll be well on your way to drawing precise, detailed, and creative fashion flats.

Adobe Illustrator workspace

Illustrator has eight main areas in its workspace:

Application bar
Use the Application bar to move, size, close, minimize, maximize and restore up or down the Illustrator screen.

Menu bar
Use the menu bar to select menu commands and open dialog boxes .

Control panel
The Control panel allows quick access to commonly used controls. The Control panel is context-sensitive which means that it changes according to the object that is selected.

Tools panel
The Tools panel contains tools that are used to create and edit objects.

Tabbed Document Window
The status bar contains a submenu for display

Artboard
The Artboard is the work area.

Panels
Panels allow quick access to commonly used controls. Additional Panels are available through the Window menu.

Status bar/ pop-up menu
The status bar contains a submenu for displaying information about the opened document. Such information includes date/time, current tool, number of available undos, document color profile, and zoom levels.

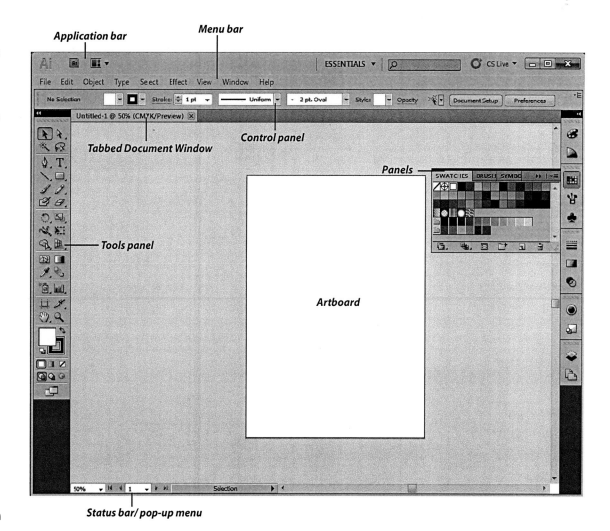

QUICK TIP!
To follow this visual chapter onscreen you may choose to launch Adobe Illustrator and create a new document. See page 42 on how to launch Adobe Illustrator.

tools panel

Adobe Illustrator houses over 75 tools in its Tools panel. To select a visible tool simply click it in the Tools panel. As you move your cursor into the Illustrator workspace it displays as the selected tool. If the Caps Lock key on your keyboard is on, the cursor will display a simple crosshair symbol. **Note:** Turning the Caps Lock key on/off toggles this feature.

The Tools panel has six categories of tools:

Selection
Tools used to select, deselect and move objects, points, paths and type. **Note:** A selected object displays a bounding box that can be used to transform (scale, rotate).

Drawing/ Type
Tools used to draw paths, objects, shapes, paintbrush strokes and type.

Transform/ Reshape
Tools used to scale, rotate, reflect, erase, warp and carve objects.

Painting
Tools used to add colors, gradients, and colorblends to objects and type. This set of tools includes the Stroke and Fill options.

Document management
Tools used to create artboards, navigate within the workspace, and zoom in and out of the artboard.

Change Screen modes
This pop-up menu allows you to change the workspace screen modes.

New to CS5

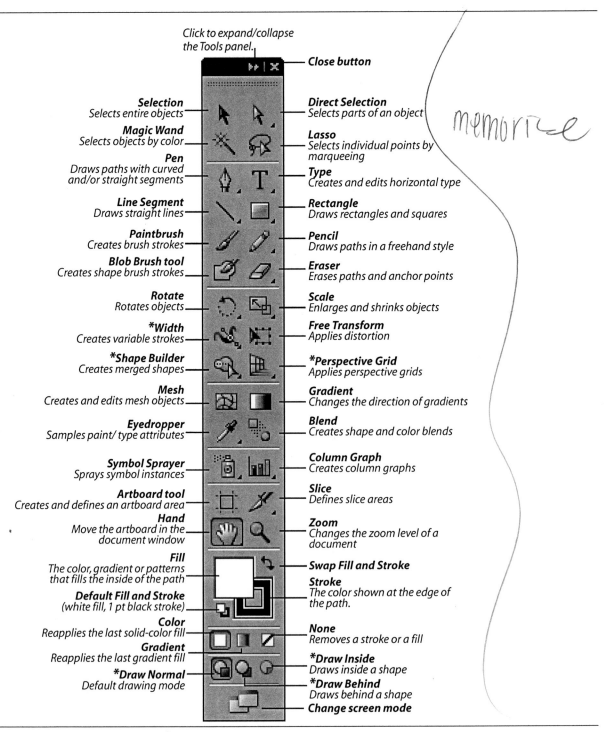

Click to expand/collapse the Tools panel.

Close button

Selection Selects entire objects

Direct Selection Selects parts of an object

Magic Wand Selects objects by color

Lasso Selects individual points by marqueeing

Pen Draws paths with curved and/or straight segments

Type Creates and edits horizontal type

Line Segment Draws straight lines

Rectangle Draws rectangles and squares

Paintbrush Creates brush strokes

Pencil Draws paths in a freehand style

Blob Brush tool Creates shape brush strokes

Eraser Erases paths and anchor points

Rotate Rotates objects

Scale Enlarges and shrinks objects

*Width Creates variable strokes

Free Transform Applies distortion

*Shape Builder Creates merged shapes

*Perspective Grid Applies perspective grids

Mesh Creates and edits mesh objects

Gradient Changes the direction of gradients

Eyedropper Samples paint/ type attributes

Blend Creates shape and color blends

Symbol Sprayer Sprays symbol instances

Column Graph Creates column graphs

Artboard tool Creates and defines an artboard area

Slice Defines slice areas

Hand Move the artboard in the document window

Zoom Changes the zoom level of a document

Fill The color, gradient or patterns that fills the inside of the path

Swap Fill and Stroke

Default Fill and Stroke (white fill, 1 pt black stroke)

Stroke The color shown at the edge of the path.

Color Reapplies the last solid-color fill

None Removes a stroke or a fill

Gradient Reapplies the last gradient fill

*Draw Inside Draws inside a shape

*Draw Normal Default drawing mode

*Draw Behind Draws behind a shape

Change screen mode

memorize

tool dialog boxes

A number of shape tools, for example the Rectangle and Ellipse, have dialog boxes that allow you to set options for creating the object. To open a shape tool dialog box click the tool in the Tools panel then click on the artboard.

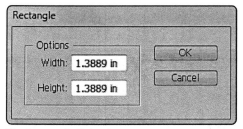

The Rectangle dialog box

Other drawing tools, for example the Pencil and Paintbrush, offer dialog boxes to set specific preferences for creating the object. To open a tool preferences dialog box double-click the tool in the Tools panel.

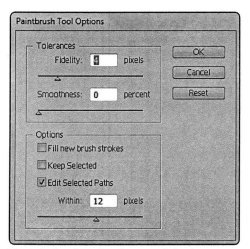

The Paintbrush Tool Preferences dialog box

Transform tools (Rotate, Reflect) also offer dialog boxes. To open a dialog box to transform an object, first select the object, then double-click a transform tool in the Tools panel.

Dialog boxes allow you to choose settings by entering a number, choosing an option from a pop-up menu, adjusting a slider or angle dial, clicking a check or button off/on or clicking a preview button off/on. The preview button allows you to view your changes while choosing the particular settings.

Double-click the Scale tool to open the Scale tool dialog box.

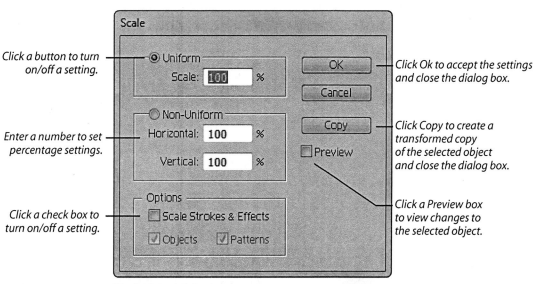

Click a button to turn on/off a setting.

Enter a number to set percentage settings.

Click a check box to turn on/off a setting.

Click Ok to accept the settings and close the dialog box.

Click Copy to create a transformed copy of the selected object and close the dialog box.

Click a Preview box to view changes to the selected object.

The Scale tool dialog box

QUICK TIP!
Illustrator's default unit of measurement is points (pt) for example, the default stroke weight in Illustrator is 1.0 pt. To use a specific unit of measurement type in a number followed by (in) for inches, (pt) for points, (p) for picas, or (mm) for millimeters.

related tools

memorize name & function (handwritten)

Each tool in the Tools panel that has a tiny corner arrow contains related tools. To select a related tool, click-hold the visible tool then drag to select the desired related tool. The related tool will then be visible in the Tools panel.

The Direct Selection related tools

You can customize the Illustrator workspace by creating a tearoff toolbar. Tearoff toolbars offer easy access to related tools. To create a tearoff toolbar click-hold then drag to the far right arrow.

The Direct Selection tearoff box

The Direct Selection tearoff toolbar

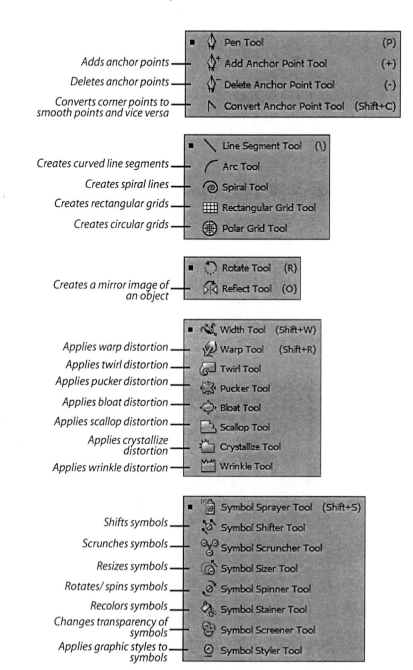

	■ Pen Tool (P)
Adds anchor points	Add Anchor Point Tool (+)
Deletes anchor points	Delete Anchor Point Tool (-)
Converts corner points to smooth points and vice versa	Convert Anchor Point Tool (Shift+C)

	■ Line Segment Tool (\)
Creates curved line segments	Arc Tool
Creates spiral lines	Spiral Tool
Creates rectangular grids	Rectangular Grid Tool
Creates circular grids	Polar Grid Tool

	■ Rotate Tool (R)
Creates a mirror image of an object	Reflect Tool (O)

	■ Width Tool (Shift+W)
Applies warp distortion	Warp Tool (Shift+R)
Applies twirl distortion	Twirl Tool
Applies pucker distortion	Pucker Tool
Applies bloat distortion	Bloat Tool
Applies scallop distortion	Scallop Tool
Applies crystallize distortion	Crystallize Tool
Applies wrinkle distortion	Wrinkle Tool

	■ Symbol Sprayer Tool (Shift+S)
Shifts symbols	Symbol Shifter Tool
Scrunches symbols	Symbol Scruncher Tool
Resizes symbols	Symbol Sizer Tool
Rotates/ spins symbols	Symbol Spinner Tool
Recolors symbols	Symbol Stainer Tool
Changes transparency of symbols	Symbol Screener Tool
Applies graphic styles to symbols	Symbol Styler Tool

related tools

mmorize — name & function

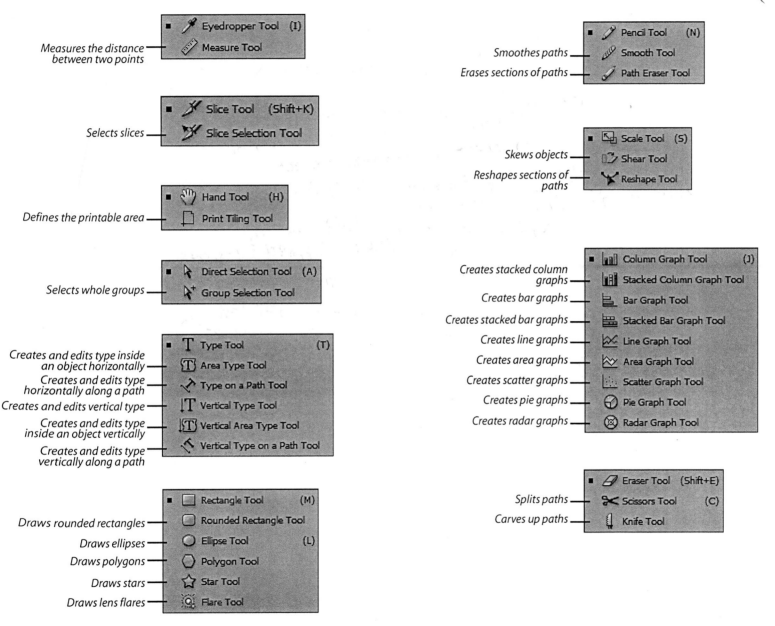

Measures the distance between two points — Eyedropper Tool (I) / Measure Tool

Selects slices — Slice Tool (Shift+K) / Slice Selection Tool

Defines the printable area — Hand Tool (H) / Print Tiling Tool

Selects whole groups — Direct Selection Tool (A) / Group Selection Tool

Creates and edits type inside an object horizontally — Type Tool (T) / Area Type Tool
Creates and edits type horizontally along a path — Type on a Path Tool
Creates and edits vertical type — Vertical Type Tool
Creates and edits type inside an object vertically — Vertical Area Type Tool
Creates and edits type vertically along a path — Vertical Type on a Path Tool

Draws rounded rectangles — Rectangle Tool (M) / Rounded Rectangle Tool
Draws ellipses — Ellipse Tool (L)
Draws polygons — Polygon Tool
Draws stars — Star Tool
Draws lens flares — Flare Tool

Smoothes paths — Pencil Tool (N) / Smooth Tool
Erases sections of paths — Path Eraser Tool

Skews objects — Scale Tool (S) / Shear Tool
Reshapes sections of paths — Reshape Tool

Creates stacked column graphs — Column Graph Tool (J) / Stacked Column Graph Tool
Creates bar graphs — Bar Graph Tool
Creates stacked bar graphs — Stacked Bar Graph Tool
Creates line graphs — Line Graph Tool
Creates area graphs — Area Graph Tool
Creates scatter graphs — Scatter Graph Tool
Creates pie graphs — Pie Graph Tool
Creates radar graphs — Radar Graph Tool

Splits paths — Eraser Tool (Shift+E) / Scissors Tool (C)
Carves up paths — Knife Tool

control panel

The Control panel at the top of the workspace is context-sensitive, meaning the controls that are shown are related to the type of object that is selected. The Control panel can be used to open a pop-up Stroke, Swatches, Character, or Transform panel.

There is a total of thirteen Control Panels:

Anchor Point
Appears when an anchor point is selected.

Blend
Appears when a blended object is selected.

Character
Appears when type characters are selected.

Envelope Warp
Appears when a warped object is selected.

Group
Appears when a group is selected.

Image
Appears when a linked, placed image is selected.

Live Paint Group
Appears when a live paint group is selected.

Mesh
Appears when a meshed or enveloped object is selected.

Mixed Objects
Appears when objects of more than one type are selected.

No Selection
Appears when there is no selection.

Path
Appears when a path, group, clipping mask, compound path, symbol, or symbol set is selected.

Tracing
Appears when a live trace object is selected.

Type
Appears when type is selected.

> **QUICK TIP!**
> The Control panel can be docked at the top or bottom of the workspace. Choose Dock to Bottom/ Dock to Top from the Control panel menu (⊙) at the far right of the panel.

The Anchor Point Control panel

selected anchor point

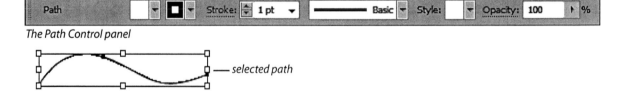

The Path Control panel

selected path

The Group Control panel

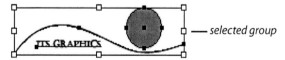
selected group

panels

Panels appear as icons and are docked along the right side of the workspace by default. To change the width of the dock, drag its gripper to the left until the panel titles appear. To expand or collapse the dock click its double arrows.

Drag to change the width of the Panels.   *Click to expand/collapse the Panels.*

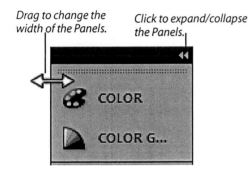

To show a panel click its icon then click its tab to show a panel within a group. To collapse a panel or panel group click the double arrow at the top right corner of the panel.

Click to collapse panel/ panel group.

Click to show panel within a group.   *Click to show panel/ panel group.*

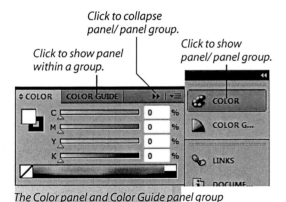

The Color panel and Color Guide panel group

Panels may be customized to fit your personal drawing needs.

dock/ undock panels
To undock a panel or panel group drag its icon or title bar outside the dock. To dock a panel drag its title bar to the right side of the workspace.

group/ ungroup panels
To create a panel group drag the panel by its tab and into the desired panel group. To ungroup a panel drag its tab away from the panel group.

panel menus
Each panel has it own set of menus. To access a panel menu click the panel's menu icon at the right side of the panel.

Expand/ Collapse

Drag the tab into/ away from a panel to group/ ungroup.   *Drag the panel by its title bar to dock/ undock.*

Close button

Show/ Hide Options

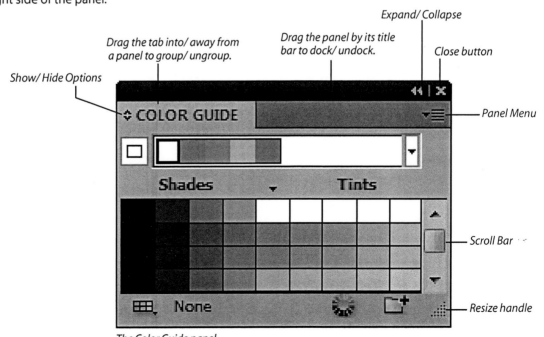

Panel Menu

Scroll Bar

Resize handle

The Color Guide panel

panels

There are more than 75 useful Panels available in Adobe Illustrator. Listed on the next two pages are the most common panels and their icons.

By default Adobe Illustrator has a basic showing of panels. To open a panel that is not showing choose it from the Window menu. In the Window menu a check mark by the panel name indicates the panel is already showing. To close a panel either choose the checked panel from the Window menu or click its close button in the panel's title bar.

New to CS5

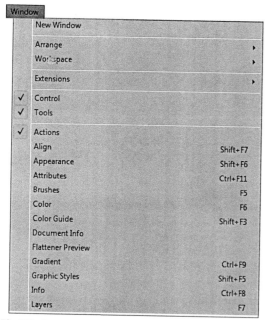

The Window menu (abbreviated)

Align
Align and/ or distribute two or more objects

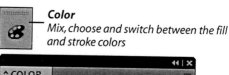

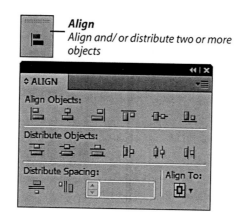

Brushes
Create and apply decorative brush strokes to paths

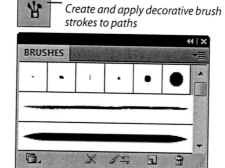

Character
Apply type attributes

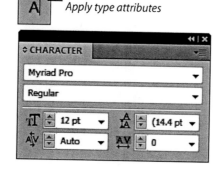

Color
Mix, choose and switch between the fill and stroke colors

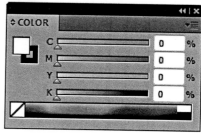

Color Guide
Manage color harmonies

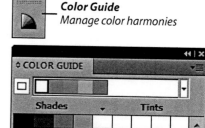

Gradient
Create, mix and apply gradients and gradient attributes

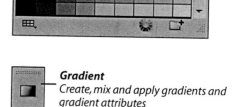

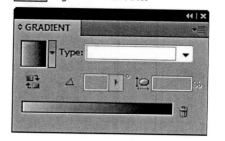

panels

Graphic Styles
Apply preset graphic styles to objects

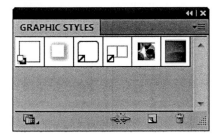

Pathfinder
Create compound shapes and cut-up shapes from multiple selected objects

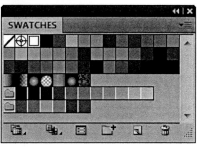

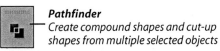

Swatches
Choose and store colors, patterns and gradients

Layers
Add and delete layers and sublayers from a document

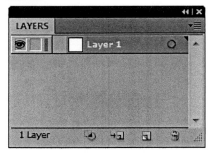

*Stroke
Edit the stroke weight, style and alignment and create dash lines Also create arrowheads and stroke width profiles.

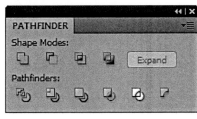

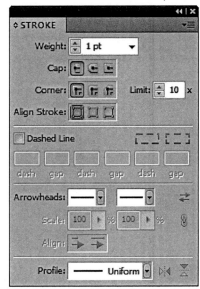

Symbols
Stores objects that can be used with the symbol sprayer tools

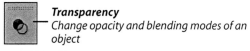

Navigator
Zoom In/Out and move the page within the document window (scroll)

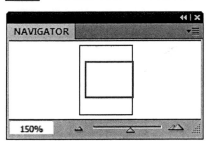

Transparency
Change opacity and blending modes of an object

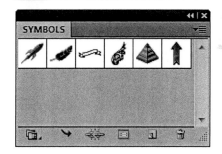

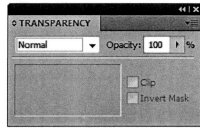

menus

The menu bar houses commands and acccess to dialog boxes.

The menu bar has ten categories of menus:

File menu
Commands for working with Illustrator files.

Edit menu
Editing of selected objects and setting Illustrator preferences.

Object menu
Commands for working with selected objects.

Type
Select font types and set font attributes using the Type menu

Select
Commands for selecting, deselecting and saving selections etc.

Effect
Apply special effects and filters to objects including drop shadows, arrowheads and zig zag distortion etc.

View
Commands for viewing, zooming and the show/ hide options for the workspace.

Window
Show/ Hide panels and navigate between several opened Illustrator documents.

Help
Access the Adobe Illustrator help center.

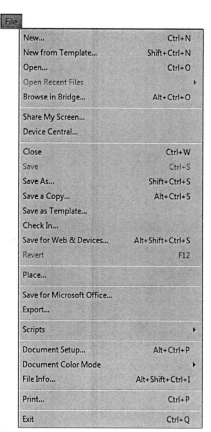

The File menu

QUICK TIP!
Note the menu command's keyboard shortcut. See page 15 for a list of commonly used Windows and Mac keyboard shortcuts.

The Edit menu

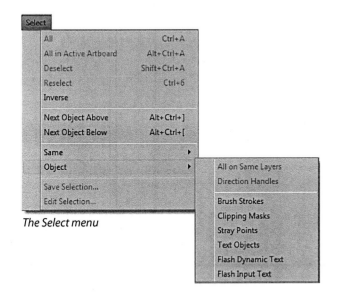

The Select menu

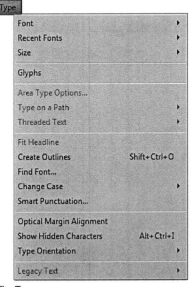

The Type menu

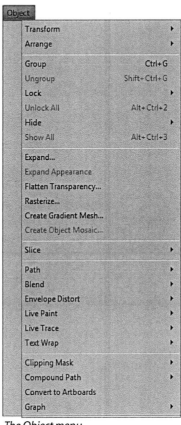

The Object menu

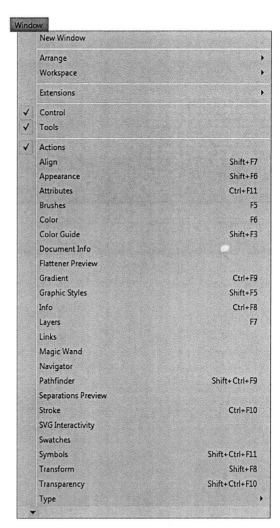

The Window menu

QUICK TIP!
A number of menu items contain sub-menus. Choose the menu command's side arrow to access its sub-menu.

view menu

Adobe Illustrator has many viewing options. Viewing options can assist in creating precise fashion flats by offering guides, rulers, grids etc.

Here are a few important View menu commands:

Zoom In/ Zoom Out
Using the view menu is one of many ways to zoom in or zoom out in Adobe Illustrator.

Show/ Hide Bounding Box
This option allows you to show or hide an object's bounding box which becomes visible when the object is selected.

Show/ Hide Print Tiling
The Show/Hide Print Tiling option allows you to show or hide the printable area on the artboard.

Show/ Hide Rulers
Rulers are helpful in measuring objects as well as aligning them. When rulers are showing you can drag from the horizontal or vertical ruler to create guides within the workspace. Also see: View > Guides when working with guides.

Show/ Hide Transparency Grid
The Show/ Hide Transparency Grid shows or hides a checkerboard background within the work area. When the transparency grid is showing you can distinguish between objects with a white fill and objects with a none fill.

Smart Guides
Smart guides are visible guides that appear when objects are moved, transformed and drawn. Its magnetic pull on objects can assist in creating precise objects.

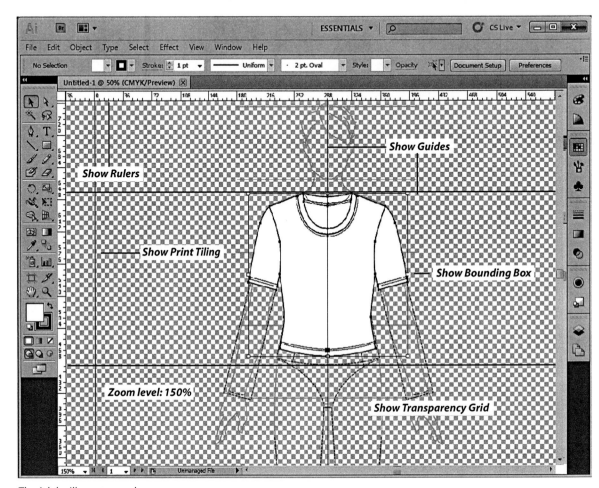

The Adobe Illustrator workspace

window tabs

When several documents are opened at one time in Adobe Illustrator their workspaces are stacked on top of one another. The Window tabs offer a convenient way to access each opened document by bringing its workspace forward in the stack. When more than one window is open click its tab to bring it forward in the stack and display its workspace.

Click the window's tab to display its workspace.

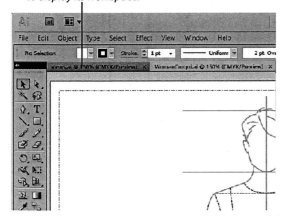

zoom in/ zoom out

Having the ability to zoom in or zoom out of the workspace is important to the drawing process. Details are easier to see when zoomed in and the entire fashion flat or artboard is visible when zoomed out.

Illustrator offers several options for zooming in/ out and scrolling the artboard:

Zoom tool
To use the Zoom tool click to select it, then click within the workspace. Hold the Alt key to toggle the zoom tool (zoom in/ zoom out).
(Mac) Option key
You can also create a zoom marquee by dragging the zoom tool across an area within the workspace.

Hand tool
Use the Hand tool to move the artboard within the workspace (scroll page). The Spacebar on your keyboard is a shortcut for the hand tool.

View menu
The View menu offers several options for zooming in and out of a workspace including Fit In Window and Actual Size.

Navigator panel
The Navigator panel is likely the most convenient way to zoom in/ zoom out and also scroll the artboard.

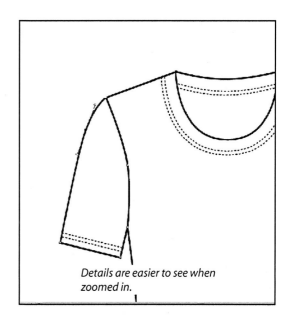

Details are easier to see when zoomed in.

Move the red box within the Navigator Preview Area to move the artboard (scroll page).

Type in a Zoom level (%).

150%

The Navigator panel

Move the Zoom Slider to zoom in/ zoom Out.

keyboard shortcuts

Keyboard shortcuts offer quick and convenient access to commonly used menu commands. Also available is the context-sensitive menu which offers shortcuts to common menu commands. Right-click on the page to open the context menu. *(Mac) Control-click*

The keyboard has eight main keys used in shortcuts:

Ctrl
The Ctrl key + a letter key allows quick access to menu commands like Copy, Paste and Undo.
(Mac) Command key + a letter key

Alt
The Alt key is used to set a transform origin and is also be used duplicate objects.
(Mac) Option key

Shift
The Shift key constrains the drawing and transforming of objects.

Tab
Use the Tab key to hide/show all panels in the workspace.

Tilde (~)
The Tilde key is used to transform a pattern fill independant of an object.

Spacebar
The spacebar is a shortcut for the Hand tool which is used for scrolling the artboard within a workspace.

Arrows (Rt, Lft, Up, Down)
Use the arrow keys to nudge/move a selected object.

Windows	*Mac*	*Result*
Ctrl+Z	*Cmd+Z*	*Undo*
Ctrl+C	*Cmd+C*	*Copy*
Ctrl+X	*Cmd+X*	*Cut*
Ctrl+V	*Cmd+V*	*Paste*
Ctrl+F	*Cmd+F*	*Paste in Front*
Ctrl+B	*Cmd+B*	*Paste in Back*
Ctrl+O	*Cmd+O*	*Open*
Ctrl+S	*Cmd+S*	*Save*
Ctrl+Q	*Cmd+Q*	*Quit*
Ctrl+(zero)	*Cmd+(zero)*	*Fit in Window*
Ctrl+1	*Cmd+1*	*Actual Size*
Ctrl+(minus)	*Cmd+(minus)*	*Zoom Out*
Ctrl +(plus)	*Cmd+(plus)*	*Zoom In*
Ctrl+A	*Cmd+A*	*Select All*
Ctrl+[+Shift	*Cmd+[+Shift*	*Send to Back*
Ctrl+]+Shift	*Cmd+]+Shift*	*Bring to Front*
Ctrl+D	*Cmd+D*	*Step Repeat*
Ctrl+J	*Cmd+J*	*Join*
Ctrl+G	*Cmd+G*	*Group*
Ctrl+Shift+G	*Cmd+Shift+G*	*Ungroup*
Alt+transform tool	*Option+transform tool*	*Set Origin/ Dialog box*
Alt+drag object	*Option+drag object*	*Duplicate*
Alt+drag handle	*Option+drag handle*	*Scale from Center*
Alt+shape tool	*Option+shape tool*	*Draw from Center*
Alt+knife tool	*Option+knife tool*	*Cut Straight*
Shift+drag	*Shift+drag*	*Constrain Object Move*
Shift+draw tool	*Shift+draw tool*	*Constrain Object Draw*
Shift+transform tool	*Shift+transform tool*	*Constrain Object Transform*
Shift+Tilde+transform tool	*Shift+Tilde+transform tool*	*Constrain Pattern Transform*
Shift+click selection	*Shift+click selection*	*Add to a Selection*
Shift+click selection	*Shift+click selection*	*Subtract from a Selection*

tool keyboard shortcuts

Utilizing the keyboard shortcuts to select tools will speed up the drawing process considerably. Practice using keyboard shortcuts as you use the lessons in this book. Also note that as you point to a tool with your cursor the tools' name and keyboard shortcut is displayed.

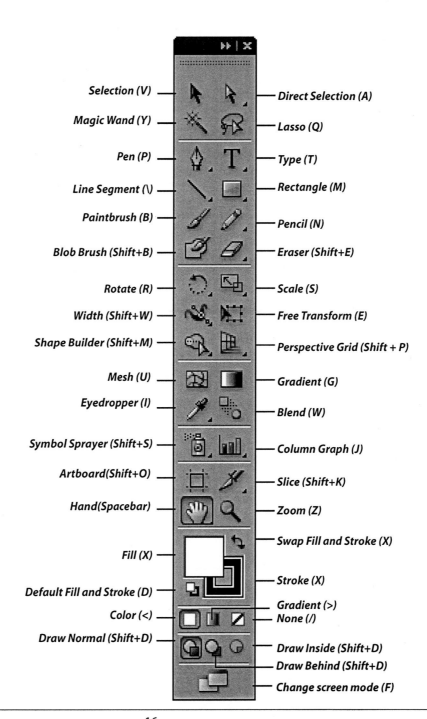

Selection (V)
Direct Selection (A)
Magic Wand (Y)
Lasso (Q)
Pen (P)
Type (T)
Line Segment (\)
Rectangle (M)
Paintbrush (B)
Pencil (N)
Blob Brush (Shift+B)
Eraser (Shift+E)
Rotate (R)
Scale (S)
Width (Shift+W)
Free Transform (E)
Shape Builder (Shift+M)
Perspective Grid (Shift + P)
Mesh (U)
Gradient (G)
Eyedropper (I)
Blend (W)
Symbol Sprayer (Shift+S)
Column Graph (J)
Artboard(Shift+O)
Slice (Shift+K)
Hand(Spacebar)
Zoom (Z)
Swap Fill and Stroke (X)
Fill (X)
Stroke (X)
Default Fill and Stroke (D)
Gradient (>)
Color (<)
None (/)
Draw Normal (Shift+D)
Draw Inside (Shift+D)
Draw Behind (Shift+D)
Change screen mode (F)

chapter two

get on your mark

Chapter two will answer this question: what exactly are you creating when you draw an object or even a fashion flat in an electronic drawing program? The answer to the question is: Adobe Illustrator uses a vector system for creating objects. A vector object is defined as a mathematically precise shape made up of anchors and paths. Anchors can be either straight or curved to create either straight or curved paths.

Get on your mark in this chapter, that is to say begin to understand and explore the anatomy of objects as they relate to fashion flats. A clear understanding of the objects that you will be creating will enhance your ability to draw precise fashion flats that are easy to reshape, transform, colorize and fabricate.

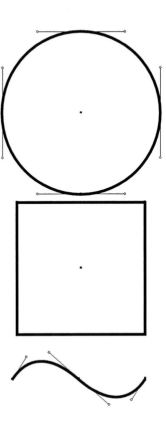

anatomy of an object

An object (shape) in Adobe Illustrator is defined by anchors, paths, direction lines, and direction points.

An object has five main parts:

Anchor point
A corner or smooth point that anchors the path.

Path
A straight or curved line between anchors.

Direction line
A line that guides the slope of a curved path.

Direction point
A handle that controls the slope of a curved path.

Note: Direction lines and direction points of a curved segment are visible when the path is selected with the Direct Selection tool.

Closed shape/ open shape
A closed shape is a continuous path with no end points. An open shape is non-continuous and has end points.

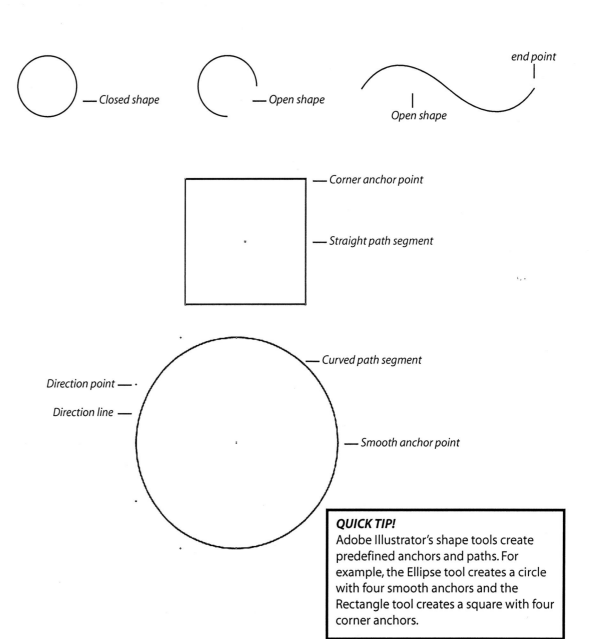

— Closed shape

— Open shape

Open shape

end point

— Corner anchor point

— Straight path segment

— Curved path segment

Direction point —

Direction line —

— Smooth anchor point

QUICK TIP!
Adobe Illustrator's shape tools create predefined anchors and paths. For example, the Ellipse tool creates a circle with four smooth anchors and the Rectangle tool creates a square with four corner anchors.

fill/ stroke

The fill of an object is the color within its outlines, the stroke of an object is the outline. Objects can have either a solid fill or a none fill (no color). To distinguish between objects that have a white fill and a none fill you may choose to show the artboard's transparency grid. To show/ hide the transparency grid choose View > Show Transparency Grid.

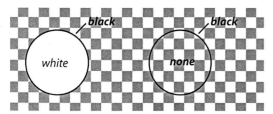

Objects as they would appear with the transparency grid showing.

A default fill and stroke defines an object or shape with a black stroke and a white fill, more specifically an object with a 1.0 pt (point) stroke weight. The main parts of your fashion flat will be closed shapes with a default fill and stroke. The fill and stroke of an object can be changed according to the detail you will create. To change the fill to a none fill simply click the None option in the Tools panel.

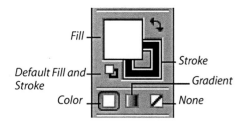

fashion flats (fill/ stroke)

When creating fashion flats it is ideal to treat its main parts as separate closed shapes (bodice, neckline, sleeves). The bodice, neckline and sleeve shapes will be drawn with a white fill and black stroke and then the fill can be easily changed once the fashion flat is complete. Open shapes such as topstitch, style lines and drape lines will be drawn with a none fill and black stroke.

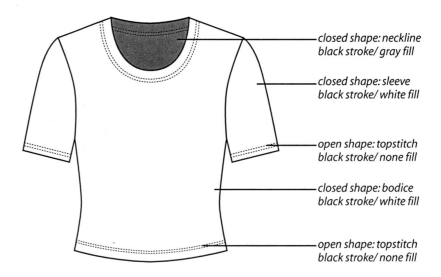

closed shape: neckline black stroke/ gray fill

closed shape: sleeve black stroke/ white fill

open shape: topstitch black stroke/ none fill

closed shape: bodice black stroke/ white fill

open shape: topstitch black stroke/ none fill

QUICK TIP!
The Stroke/ Fill option in the Tools panel will show the stroke and fill attributes of a selected object. A question mark appears when several objects with varying attributes are selected.

stroke weight

By default an object's stroke weight is 1.0 pt (point). The stroke weight of an object can be increased to appear as a thicker outline and/or a dash line attribute can be applied (as it would appear in topstitching). The stroke weight can be set before or after the shape is created.

— weight: 1.0 pt   — weight: 5.0 pt

Use the Stroke panel's pop-up menu in the Control panel to set an object's stroke Weight, Dashed Line attributes, add Arrowheads and view the stroke's Profile.

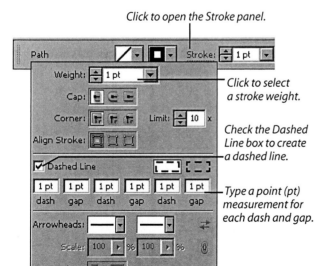

Click to open the Stroke panel.

Click to select a stroke weight.

Check the Dashed Line box to create a dashed line.

Type a point (pt) measurement for each dash and gap.

fashion flats (stroke weight)

You will use a variation of stroke weights and attributes when drawing fashion flats depending on the effects that you will want to create. An idea to keep in mind is closed shapes (bodice, neckline, sleeves) will be drawn with a stroke weight of 1.0 pt. Topstitch, drape lines and small details are best when drawn with a stroke weight of 0.5 pt.

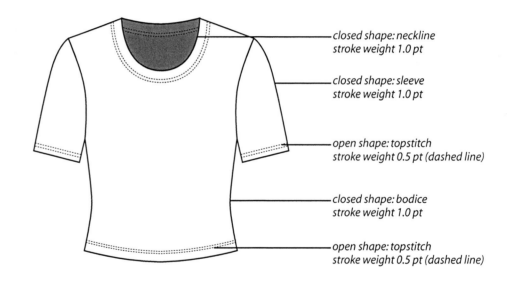

closed shape: neckline
stroke weight 1.0 pt

closed shape: sleeve
stroke weight 1.0 pt

open shape: topstitch
stroke weight 0.5 pt (dashed line)

closed shape: bodice
stroke weight 1.0 pt

open shape: topstitch
stroke weight 0.5 pt (dashed line)

stacking order

Each time an object is drawn or pasted into a workspace it is stacked on top of existing objects. For example, if you draw a circle and then draw a rectangle the rectangle is stacked on top of the circle. This is important to note when creating fashion flats because often you will create objects that stack either on top of or beneath another object.

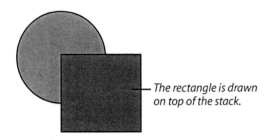

— The rectangle is drawn on top of the stack.

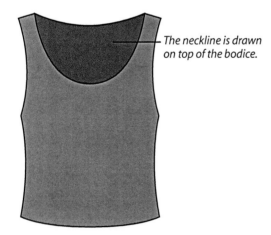

— The neckline is drawn on top of the bodice.

fashion flats (stacking order)

When drawing a fashion flat you will start by drawing the bodice shape and then you will draw the bodice style lines, topstitch and drape lines on top of the bodice shape. Next you will create the neckline shape then draw its topstitch on top of the neckline shape. Complete the fashion flat with the sleeve shapes then draw its topstitch on top of the sleeve shape. The sleeves are then sent beneath the bodice so as to maintain the bodice armhole shape.

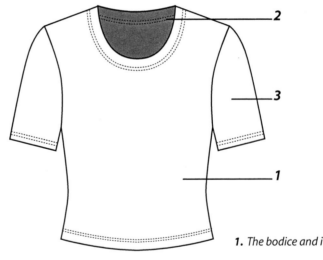

1. The bodice and its topstitch is drawn.

2. The neckline and its topstitch is drawn on top of the bodice.

3. The sleeve and its topstitch is drawn on top of the bodice.

bring to front/ send to back

fashion flats (stacking order)

Once an object is created on top of the stack it can then be sent to the back of the stack if needed. Essentially you can change the stacking order using the Bring to Front/ Send to Back command from either the Object > Arrange menu or by using the context menu. To use the context menu right-click on the artboard. *(Mac) Control-click*

Certainly you can control the order in which you create each part of the fashion flat. For example you will draw the bodice first and then the neckline. This will conveniently stack the neckline on top of the bodice. Lets say you create the neckline first and then the bodice. To change the stacking order select the bodice and choose Object > Arrange > Send to Back.

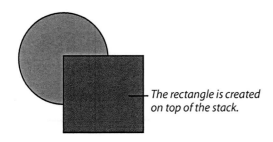

The rectangle is created on top of the stack.

Choose Object > Arrange > Bring to Front or Send to Back.

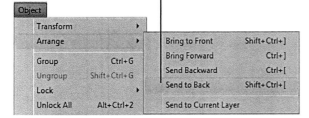

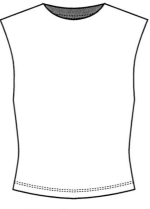

The bodice is created on top of the neckline.

The bodice is sent beneath the neckline.

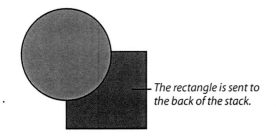

The rectangle is sent to the back of the stack.

anatomy of a fashion flat

Efficiency is the goal when drawing fashion flats in an electronic drawing program.

Fashion designers look to explore design options in the form of bodice shapes (fitted, unfitted, a line, cropped) and details (scoop neckline, v neckline).

Creating proportioned fashion flats, using a standard croqui, where shapes and details are interchangeable like "puzzle pieces" is the key to efficient design.

grouped shapes and details

The bodice and all its parts are grouped to make one component of the fashion flat. The neckline and its parts are grouped and then the sleeve and its parts are grouped into one component. Developing a habit of grouping the seperate parts of your fashion flats will benefit you in the end. Doing this will make exploring the various fashion ideas easier and more efficient.

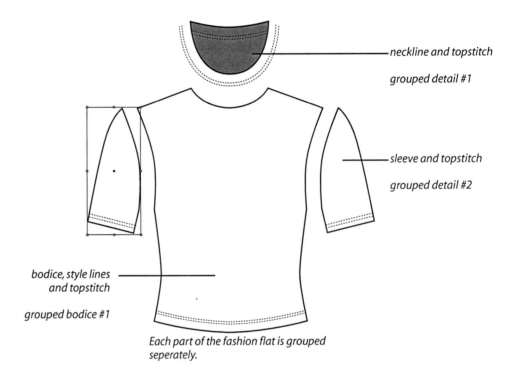

neckline and topstitch

grouped detail #1

sleeve and topstitch

grouped detail #2

bodice, style lines and topstitch

grouped bodice #1

Each part of the fashion flat is grouped seperately.

A grouped fashion flat.

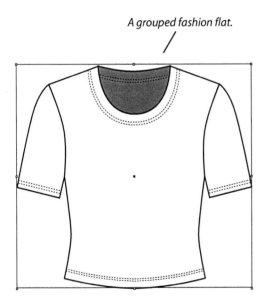

SnapFashun.com

Full credit to the "puzzle pieces" concept belong to SnapFashun Inc, creators of **The SnapFashun Library** (www.snapfashun.com). A company who, from the very first introduction of electric drawing programs, created an extensive library of interchangeable items (bodice shapes) and details (necklines, sleeves etc.) that "snap" together like puzzle pieces, in an infinite array of designs. The SnapFashun Library, the only extensive library of its kind in the fashion industry today, continues to inspire fashion designers around the globe with their online trend service.

explore design

By adopting SnapFashun drawing techniques a designer can explore designs from a simple fashion flat, create a duplicate of the original fashion flat and then explore new necklines, sleeves, style lines or topstitch.

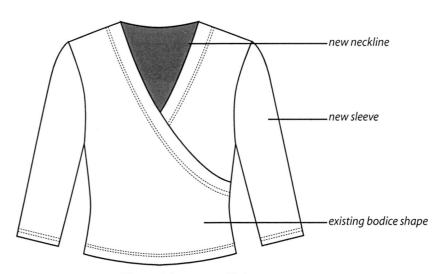

new neckline

new sleeve

existing bodice shape

A new neckline and sleeves are added to the existing bodice.

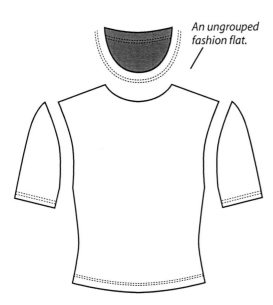

An ungrouped fashion flat.

chapter three

get ready
beginner level lesson

This is an exciting chapter where you will learn how to draw and edit objects using Adobe Illustrator's tools. You'll also explore reshaping, transforming, and grouping objects all to prepare for drawing and editing your fashion flats.

Launch Adobe Illustrator and create a new document so that you can practice the techniques demonstrated in this chapter.

Located on the Electric Fashion CD is a folder called Draw Practice. In the Draw Practice folder you will find Adobe Illustrator documents with usefull practice lessons. Give the practice lessons a spin and get ready to be amazed by your ability to create objects with great precision and speed.

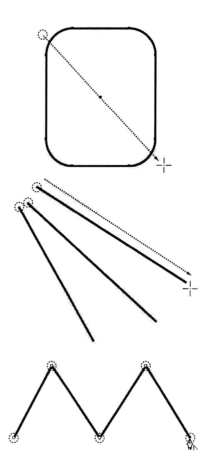

undo/ redo

The Undo/ Redo command is the greatest thing since sliced bread. To undo the last step choose Edit > Undo. Repeat Edit > Undo to undo up to 200 times. To reverse the undo choose Edit > Redo.

You can undo or redo while a document is open and even after you save it but not after the document is closed and re-opened.

Choose Edit > Undo to undo an action.

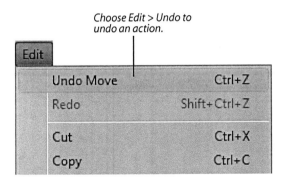

Choose Edit > Redo to reverse an action.

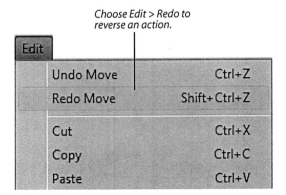

context menu

An even easier way to choose a command is by using the onscreen context-sensitive menu. To open the context menu to undo or redo an action simply right-click on the artboard. *(Mac) Control-click*

Right-click on the artboard to open the context menu.

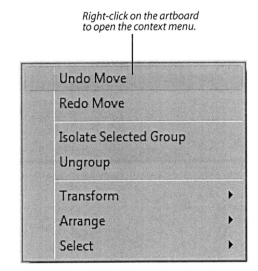

QUICK TIP!
A convenient way to undo/redo is by using the undo keyboard shortcut. Press **Ctrl+Z** on your keyboard to undo an action. Press **Shift+Ctrl+Z** on your keyboard to redo. *(Mac) Command+Z and Shift+Command+Z*

---

shape tools

Shape tools create predefined anchors and paths. For example, the Ellipse tool creates circles with four smooth anchor points.

Before you begin drawing click the Default Fill and Stroke option located in the Tools panel. This will ensure that the object you create has a white fill and a black stroke.

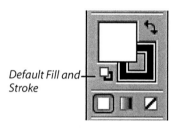

Default Fill and Stroke

There are five simple shapes commonly used in drawing fashion flats:

Rectangle
Ideal for creating pockets, tabs, buttonholes, zipper pulls and zipper casings.

Rounded Rectangle
Ideal for creating rounded pockets and various rounded rectangular details.

Ellipse
Use this tool to create buttons, grommets, rivets and various circular details.

Polygon
Use this tool to create novelty polygonal shapes for buttons, pockets and patches.

Star
Use this tool to create novelty star shapes, appliques, logos and star graphics.

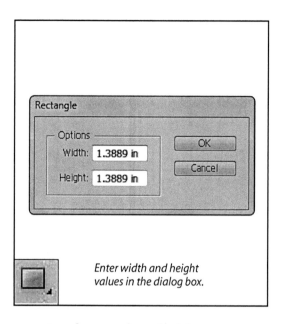

Enter width and height values in the dialog box.

create shapes by clicking

1. Select a shape tool (**Rectangle or Ellipse**).

2. Click on the page to set the origin of the shape. A shape dialog box opens. Type in measurements for your shape and click **OK**.

   Note: Be certain to specify the desired units of measurement. For inches type a number then "in". For points type a number and "pt".

3. Select the **Selection** tool to cancel the shape tool.

4. Click outside the object (on the page) to deselect.

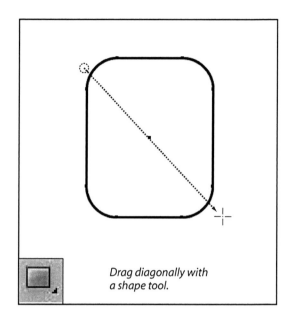

Drag diagonally with a shape tool.

create shapes by dragging

1. Select a shape tool (**Rectangle or Ellipse**).

2. Point to an area on the page then drag diagonally until the shape is the desired size.

3. Select the **Selection** tool to cancel the shape tool.

4. Click outside the object (on the page) to deselect.

QUICK TIP!
To create a shape from its center origin hold the **Alt key** while creating the shape. *(Mac) Option key*
To create a proportioned shape (square/circle) hold the **Shift key** while creating the shape.

---

freehand tools

Drawing freehand in Adobe Illustrator is an option for creating style lines, drape lines, details and graphics for your fashion flats.

These three tools allow you to create paths using a freehand technique:

Line Segment
Use this tool to create straight lines (one corner anchor at each end). **Note:** By default the line segment tool creates lines with a none fill and a black stroke.

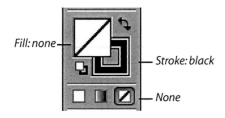

Fill: none — *Stroke: black* — *None*

Pencil
A freehand drawing tool that allows you to create pencil strokes.

Paintbrush
A freehand drawing tool that allows you to create brush strokes. Use the Brushes panel to choose a brush style.

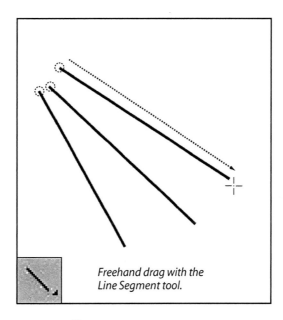

Freehand drag with the Line Segment tool.

create lines

1. Select the **Line Segment** tool.

2. Point to an area on the page then drag to create a line.

3. To create more lines repeat step 2.

4. Select the **Selection** tool to cancel the **Line Segment** tool.

5. Click outside the object (on the page) to deselect.

   Note: To create a perfect horizontal, vertical or 90 degree angle line hold the **Shift key** while drawing the line.

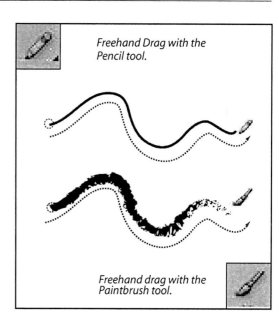

Freehand Drag with the Pencil tool.

Freehand drag with the Paintbrush tool.

create pencil strokes

1. Select the **Pencil** tool.

2. Point to an area on the page then drag to create a pencil stroke.

3. With the **Selection** tool click outside the object (on the page) tool to deselect.

create brush strokes

1. Select a brush style from the **Brushes** panel.

2. Select the **Paintbrush** tool.

3. Point to an area on the page then drag to create a paintbrush stroke.

4. With the **Selection** tool click outside the object (on the page) tool to deselect.

pen tool/ width tool

Pen
This tool is unique because it allows you to set anchors exactly where you want to create either straight or curved paths. The Pen tool is the primary tool used to create fashion flat shapes and complex details.

As you use the Pen tool to create shapes note these 10 Pen symbols: **Note:** Be certain the Caps Lock key on your keyboard is off.

New path

Creating a path

Connecting a path

Subtracting an anchor point

Adding an anchor point

Closing a path

Connecting two paths

Connecting a selected path

Creating a curved

Width
The Width tool allows you to vary the width on a path. Once you have adjusted the width you can then save custom width profiles, which can be applied to any stroke. See the Variable Width Profile in the Control Panel.

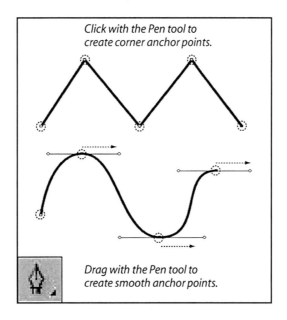

Click with the Pen tool to create corner anchor points.

Drag with the Pen tool to create smooth anchor points.

create straight paths

1. Select the **Pen** tool.

2. Point to an area on the page then click to set the first anchor.

3. Release and reposition then click to create a second anchor. Repeat to create additional anchors.

create curved paths

1. Select the **Pen** tool.

2. Click on the page to set the first anchor.

4. Release and reposition then drag to create a smooth anchor. Repeat to create additional anchors either straight or curved.

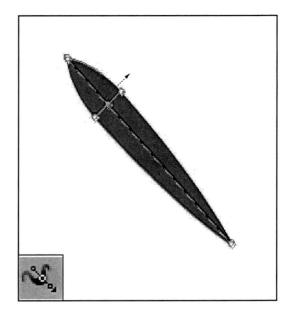

create varied widths

1. Select the **Width** tool.

2. Point to a section on a path then drag to create width.

select/ deselect/ direct select

Selecting an object is a way of identifying what will be edited (reshaped or transformed). Deselecting an object (clicking on the page away from all objects) is a way of identifying that no shape will be edited.

Adobe Illustrator has six main selection tools:

Selection
Used to select, move, scale, rotate or reflect using the object's bounding box. Click outside the object (on the page) with the selection tool to deselect. The selection tool is also used to drag a selection marquee over the object to select it.

Direct Selection
Used to select or move an anchor or direction point. Reshape an object by moving its anchors or its direction points.

Group Selection
Click once within a group to select a single object and edit it individually. A second click will select the next object within the group and so on.

Magic Wand
Used to select shapes of same or similar fill and stroke attributes.

Lasso
Use this selection tool to select anchors with a freehand selection marquee.

Shift+click
To select more than one object hold the Shift key on your keyboard then Shift+click the other objects to include them in the selection. You can also use shift+click to release an object from a selection.

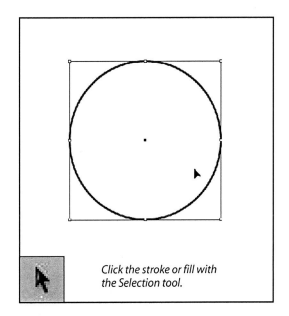

Click the stroke or fill with the Selection tool.

select

1. Select the **Selection** tool.

2. Click the stroke or fill (if it has a fill) of the object to select it.

   Note: You'll see the objects' highlighted bounding box.

OR

2. Position the **Selection** tool outside the object then drag diagonally a selection marquee across the object.

   Note: All objects within the selection marquee will be selected.

deselect

1. With the **Selection** tool click outside the object to deselect.

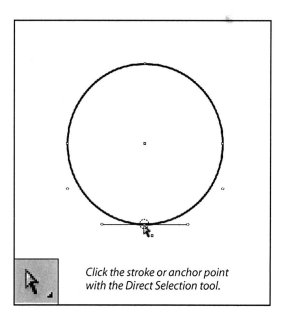

Click the stroke or anchor point with the Direct Selection tool.

direct select

1. Select the **Direct Selection** tool.

2. Click the stroke or an anchor on the object (not the fill).

   Note: The Direct Selection arrow will display a white box next to it when are pointing to an anchor or path.

reshape

Anchor points and path segments can be reshaped using various reshape tools and commands. Most commonly the Direct Selection tool is used to reshape an object. Become familiar with how the object appears when it is selected with the Direct Selection tool. When an object is selected with the Selection tool you will see its bounding box. When an object is selected with the Direct Selection tool you will see its anchors, paths and direction points.

Note the four most common tools used to reshape an object.

Direct Selection
Used to select and move an anchor point, path or direction point. The direction point controls the slope of a curve.

Add Anchor Point
Adds an anchor to a path segment.

Delete Anchor Point
Deletes an anchor from a path segment.

Convert Anchor Point
Converts a corner anchor into a smooth anchor and vice versa.

> **QUICK TIP!**
> Explore the Anchor Point **Control Panel** (visible when an anchor is selected) for quick reshaping commands.

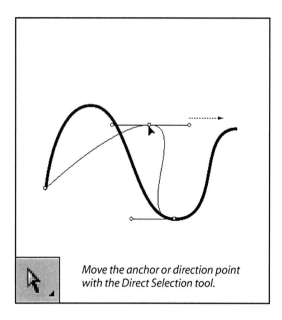

Move the anchor or direction point with the Direct Selection tool.

reshape

1. Select the **Direct Selection** tool.

2. Click the stroke of the object (not the fill).

3. Click directly on top of an anchor (only the selected anchor will be highlighted).

4. Drag (move) the anchor to reshape the object.

5. Drag (move) the direction point (shown only when selecting a curved path) to reshape the path.

add anchor point

1. Select the **Add Anchor Point** tool.

2. Click on the stroke of the object.

3. Use the **Direct Selection** and/or the **Convert Anchor Point** tool to reshape the object.

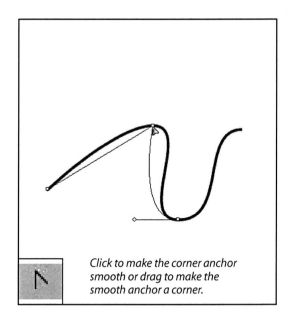

Click to make the corner anchor smooth or drag to make the smooth anchor a corner.

convert anchor point

1. Select the object with the **Selection** tool.

2. Select the **Convert Anchor Point** tool.

3. Click a smooth anchor point to make it a corner anchor point.

   OR

3. Drag a corner anchor point to convert it into a smooth anchor.

4. Use the **Direct Selection** to reshape the object.

reshape

Adobe Illustrator's Tools panel offers a number of tools for reshaping including the Pencil tool, Smooth tool, the Path Eraser tool, Reshape tool and the collection of Warp tools. Be certain to explore them all as you may discover unique uses for each of them.

Note the three most common reshape tools used to reshape an object.

Eraser
The Eraser is a reshaping tool that allows erase parts of a shape. Choose an eraser size and shape by double-clicking the Eraser tool in the Tools panel.

Scissors
Used to split a path segment. Use the Selection tool to seperate the cut pieces.

Knife
Used to carve an object. This tool can be used freehand or hold the Alt and Shift keys to constrain the carve.
(Mac) Option key and Shift key

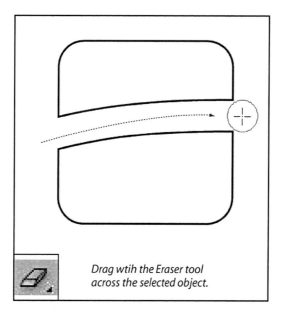

Drag wtih the Eraser tool across the selected object.

erase

1. Select the object with the **Selection** tool.

2. Select the **Eraser** tool.

3. Double-click the **Eraser** tool to set the eraser's angle, roundness and diameter. Click **OK**.

4. Drag the eraser accross the object to erase a section of the object.

   Note: The eraser will erase several objects unless an individual object is specified by a selection.

5. With the **Selection** tool click outside the object to deselect.

   Note: To erase in a perfect horizontal, vertical or 90 degree angle hold the **Shift key** while erasing.

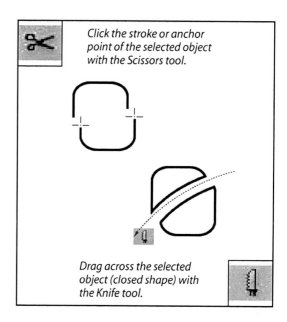

Click the stroke or anchor point of the selected object with the Scissors tool.

Drag across the selected object (closed shape) with the Knife tool.

scissors

1. Select the **Scissors** tool.

2. Click the stroke or an anchor point on the object. **Note:** Closed shapes will be in cut in two places.

3. Select and move the cut object with the **Selection** tool.

knife

1. Select the object with the **Selection** tool.

2. Select the **Knife** tool.

3. Start outside the object then drag across to the opposite outer side.

4. Deselect the object with the **Selection** tool then select and move the carved object.

transform

The simplest way to move, scale or rotate an object is by using its bounding box (visible when selected). Also, the Tools panel offers a selection of transform tools for scaling, rotating, reflecting and shearing objects. Each transform tool has a dialog box. To open its dialog box double-click the transform tool in the Tools panel.

Note the five most common transforming options:

Move
Change the position of the object on the page (click to select the object then drag to move it).

Scale
Resize the object using its bounding box or use the Scale tool dialog box to type in a percentage increase/decrease for the object, its pattern, its stroke weight and/or to create a scaled copy.

Rotate
Rotate the object using its bounding box or use the Rotate tool dialog box to type in an angle degree for the object, its pattern and/or to create a rotated copy.

Reflect
Use the Reflect tool to set a reflect origin (Alt+click) then use the Reflect dialog box to set the reflect axis and angle. Click Copy to create a mirror image of the object.

Shear
Use the Shear tool dialog box to choose an axis and type in an angle degree for the object, its pattern and/or to create a sheared copy.

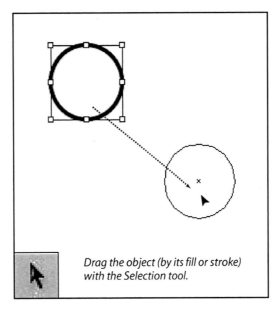

Drag the object (by its fill or stroke) with the Selection tool.

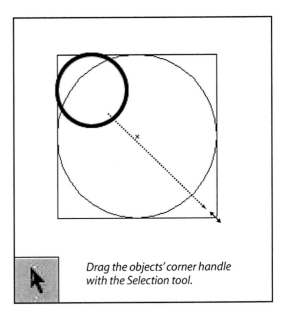

Drag the objects' corner handle with the Selection tool.

move

1. Select the object with the **Selection** tool.

2. Drag the object by its fill or stroke to move it.

   Note: Hold the **Shift key** while moving to constrain the movement horizontally, vertically or at a 45° angle.

duplicate

1. Select the object with the **Selection** tool.

2. Hold the **Alt key** on your keyboard and drag (move) the object to duplicate it.

   Note: To move and duplicate in a perfect horizontal, vertical or 45° angle hold the **Alt** and **Shift keys** while moving the object. *(Mac) Command + Shift*

scale

1. Select the object with the **Selection** tool.

2. Position the pointer on the corner handle until you see the double-head scale arrow.

3. Drag the corner handle to resize the object.

   Note: Hold the **Shift key** while scaling to constrain the proportions.

OR

2. Double-click the **Scale** tool to open its dialog box.

3. Check the **Preview** box.

4. Type in a number (0%) for the scale.

5. Click **OK** to scale the object or click **Copy** to create a scaled copy of the object.

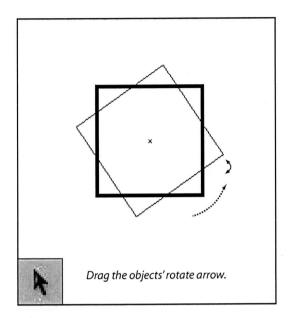

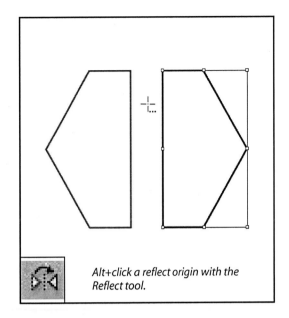

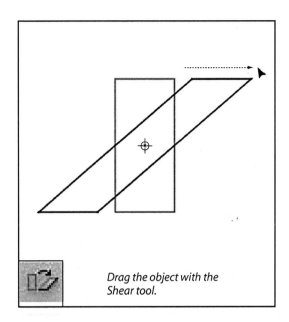

Drag the objects' rotate arrow.

Alt+click a reflect origin with the Reflect tool.

Drag the object with the Shear tool.

rotate

1. Select the object with the **Selection** tool.

2. Position the pointer just outside the corner handle until you see the double-head rotate arrow.

3. Drag the rotate arrow to rotate the object.

OR

2. Double-click the **Rotate** tool to open its dialog box.

3. Check the **Preview** box.

4. Type in a number (0°) for the rotate.

5. Click **OK** to rotate the object or click **Copy** to create a rotated copy of the object.

6. Select the **Selection** then click outside the object to deselect.

reflect

1. Select the object with the **Selection** tool.

2. Select the **Reflect** tool.

3. Hold the **Alt key** on your keyboard and click on the page at the desired reflect origin.

4. In the reflect dialog box click the **Preview** button.

5. Click the desired axis and type in a number (0°) for the reflect.

6. Click **OK** to reflect the object or click **Copy** to create a reflected copy of the object.

7. Select the **Selection** tool then click outside the object to deselect.

shear

1. Select the object with the **Selection** tool.

2. Select the **Shear** tool.

3. Drag horizontally or vertically to shear.

OR

2. Double-click the **Shear** tool to open its dialog box.

3. Check the **Preview** box.

4. Type in a number (0°) for the shear angle and choose an axis angle.

5. Click **OK** to shear the object or click **Copy** to create a sheared copy of the object.

6. Select the **Selection** then click outside the object to deselect.

copy/ paste

The Copy/ Paste command is used to make a copy of an object. This command is an ideal way to copy objects from one document to another. To copy an object (first select it) then choose Edit > Copy from the menu bar. The object is copied to the clipboard (temporary holding place). To paste a copied object choose Edit > Paste from the menu bar.

Choose Edit > Copy to copy an object.

Edit	
Undo Pen	Ctrl+Z
Redo	Shift+Ctrl+Z
Cut	Ctrl+X
Copy	Ctrl+C
Paste	Ctrl+V

Choose Edit > Paste to paste a copied object.

Edit	
Undo Pen	Ctrl+Z
Redo	Shift+Ctrl+Z
Cut	Ctrl+X
Copy	Ctrl+C
Paste	Ctrl+V

paste in front/ paste in back

The Paste in Front/ Paste in Back command allows you to paste an object exactly from where it was copied on the page, either on top of or behind the original object. This command is used to copy an object or path then paste the object or path in the exact position from where it was copied.

Choose Edit > Paste in Front or Paste in Back.

Edit	
Undo Pen	Ctrl+Z
Redo	Shift+Ctrl+Z
Cut	Ctrl+X
Copy	Ctrl+C
Paste	Ctrl+V
Paste in Front	Ctrl+F
Paste in Back	Ctrl+B

> **QUICK TIP!**
> A convenient way to copy an object is by using a keyboard shortcut. Press **Ctrl+C** on your keyboard to copy an object. Press **Ctrl+V** on your keyboard to paste an object. *(Mac) Command+C and Command +V*

align/ distribute

The Align Panel offers options for aligning and distributing two or more objects. A good use for this panel is aligning and evenly distributing several buttons. **Note:** The buttons must be individually grouped.

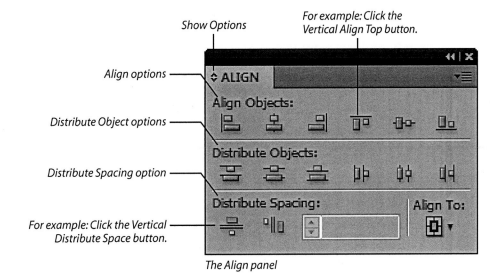

Show Options

For example: Click the Vertical Align Top button.

Align options

Distribute Object options

Distribute Spacing option

For example: Click the Vertical Distribute Space button.

The Align panel

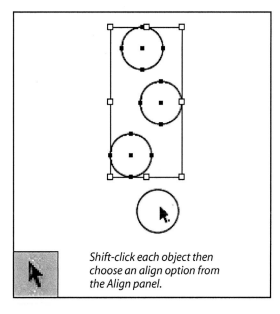

Shift-click each object then choose an align option from the Align panel.

align

1. Select the **Selection** tool.

2. Hold the **Shift key** on the keyboard.

3. **Shift-click** each object that will be aligned.

   Note: This will constrain the selection allowing you to select several objects.

4. Release the Shift key then click an align button in the **Align** panel.

distribute

5. With the objects still selected click a distribute spacing button from the **Align** panel.

offset path

The Offset Path command creates a duplicate outline of a selected object either inside (- before the number) or outside (+ before the number). An ideal unit of measurement for this command is points (pt) as it also applies to the measurment for stroke weight (default stroke weight = 1 pt).

A good use for this command is when creating ties or strings, binding, waistbands, belts and topstitching detail on pockets.

If the shape is a closed shape the offset path will create a duplicate shape either inside or outside the original shape. If the shape is an open shape the offset path will create an outline of the original path or line.

Type in a number (+/-) followed by (pt) for the amount of the offset.

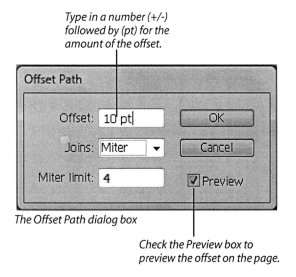

The Offset Path dialog box

Check the Preview box to preview the offset on the page.

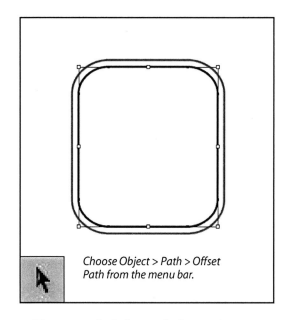

Choose Object > Path > Offset Path from the menu bar.

offset path (closed shape)

1. Select a closed shape with the **Selection** tool.

2. Choose **Object > Path > Offset Path** from the menu bar.

3. Check the **Preview** box.

4. Type in a number for the **Offset** and include (pt) after the number for example (**-3.5pt**).

5. Click **OK**.

6. Select the **Selection** tool then deselect the shapes.

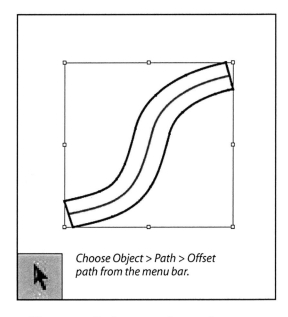

Choose Object > Path > Offset path from the menu bar.

offset path (open shape)

1. Select an open shape with the **Selection** tool.

2. Choose **Object > Path > Offset Path** from the menu bar.

3. Click the **Preview** button.

4. Type in a number for the **Offset** and include (pt) after the number for example (**7.0pt**).

5. Click **OK**.

6. Select the **Selection** tool then deselect the shapes.

7. Use the **Selection** tool to move the desired shape.

make with warp

The Make with Warp command offers options for applying a warp to a selected object. There are many uses for this command such as creating waistbands, necklines, pocket and warped type.

The Make with Warp command offers an array of warp styles, the option to bend horizontally or vertically and the option to distort the warp.

expand!

Important Note: Be certain to expand the object once you have completed the warp command. Doing this will set the warp and convert the object into anchors and paths that can then be reshaped. To do this choose Object > Expand from the menu bar then click OK in the Expand dialog box.

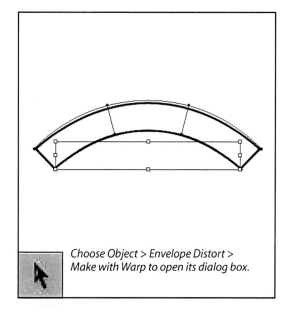

Choose Object > Envelope Distort > Make with Warp to open its dialog box.

Click the downward arrow to choose a warp style.

Click the Preview box to preview the warp.

Choose either a horizontal or vertical angle for the warp.

Move the Bend slider to adjust the bend amount (%)

Move the Horizontal slider to adjust the distort amount.

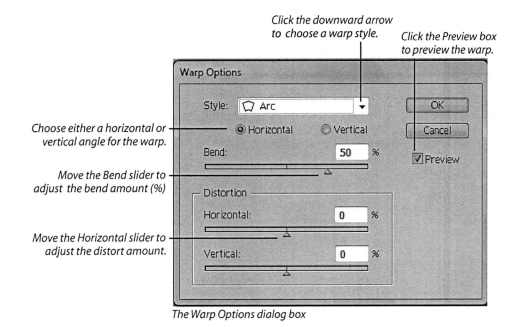

The Warp Options dialog box

make with warp

1. Select an object with the **Selection** tool.

2. Choose **Object > Envelope Distort > Make with Warp** from the menu bar.

3. Check the **Preview** box.

4. Choose a **Style** for the warp and select **Horizontal** or **Vertical**.

5. Drag the **Bend** slider or type in a number (0%) for the amount and direction of the bend.

6. Click **OK**.

expand!

1. Choose **Object > Expand** then click **OK** to complete the warp.

unite

The Pathfinder panel is used to create a compound object from two or more shapes. When fashion flats are created the initial shape is the half bodice shape. The half shape is then copy-reflected to the other side using the Reflect tool. Finally, the Unite option in the Pathfinder panel is used to combine the two halves to create one complete closed shape. Combining the two sides will make the object easier to edit, colorize and fabricate.

CS3 expand!

Important Note: If using earlier versions of Adobe Illustrator (CS2 or CS3) be certain to click the Expand button in the Pathfinder panel once the shapes are combined.

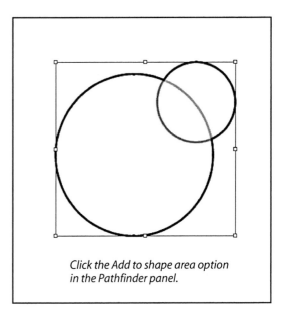

Click the Add to shape area option in the Pathfinder panel.

unite

1. **Shift+click** two or more overlapping objects with the Selection tool.

2. Click the **Unite** button in the **Pathfinder** panel.

Shape Mode options ——

For example: Click the Unite —— button.

Pathfinders options. ——

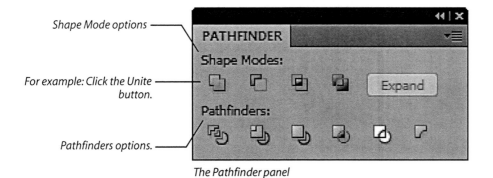

The Pathfinder panel

QUICK TIP!
Your cursor displays the title for each Pathfinder option when you point to them in the Pathfinder panel.

group/ ungroup

Grouping objects is essentially like glueing all its parts together thus making it easier to select, move, duplicate, copy and transform the group as one unit. Grouped objects can be ungrouped and seperated so that you can edit the parts individually. Also, grouped objects can be edited while remaining in the group by using the Group Selection tool or the Isolated Group command. When objects are grouped they share a bounding box (visible when selected). When objects are ungrouped each object has its own bounding box.

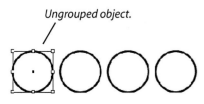

Grouped objects.

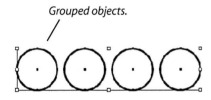

Ungrouped object.

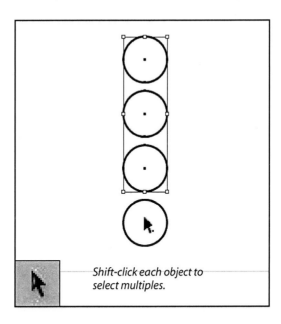

Shift-click each object to select multiples.

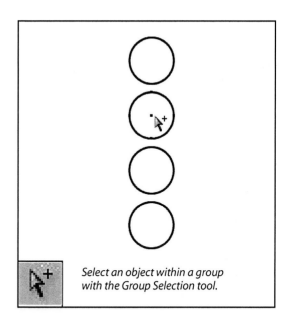

Select an object within a group with the Group Selection tool.

group

1. **Shift-click** each object that will be in the group with the **Selection** tool.

2. Choose **Object > Group** from the menu bar.

ungroup

1. Select the grouped objects with the **Selection** tool.

2. Choose **Object > Ungroup**.

3. Deselect the object with the **Selection** tool.

4. Use the **Selection** tool to edit and/or transform the individual object.

group selection

1. Select an object within a group with the **Group Selection** tool.

2. Edit and/or transform the selected object.

isolated group

1. **Double-click** the grouped objects with the **Selection** tool.

   Note: A gray bar will appear at top of the workspace indicating the **Group** isolation mode.

2. Select and edit and/or transform objects within the group.

3. Click the **Group isolation mode** side arrow to exit isolation mode.

QUICK TIP!

In isolation mode objects in an isolated group are editable and all other objects are dimmed. To enter isolation mode double-click the grouped objects. To exit isolation mode, click the gray bar at the top of the workspace.

chapter four

get setup and go

In this chapter you will launch Adobe Illustrator, and open the women's croqui document from the Electric Fashion CD. You'll then setup and customize your workspace using the View and Window menu.

The croqui in this document exists on Layer 1 which has been renamed croqui. In this tutorial you will create a new layer to be used like a piece of tracing paper over the croqui layer. Doing this will allow you to create your fashion flats on one layer using the croqui as your guide on the bottom layer.

It is ideal to setup your workspace before you go to draw each fashion flat in the following chapters. Doing so will make for a familiar workspace and easy navigation.

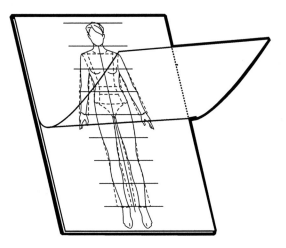

launch Adobe Illustrator

Not all computers are setup identically but often times on a Windows computer you can launch Adobe Illustator from either the desktop or from the Start button found on the Window's taskbar.

Window's Start button

Macintosh computers can vary slightly depending on the operating system (OS). To launch Illustrator on a Mac choose Adobe Illustrator from the Mac OS dock at the bottom of the screen.

Mac OS menu dock.

Double click the Adobe Illustrator icon on the desktop.

launch Adobe Illustrator (win)

1. Double-click the **Adobe Illustrator** icon located on the desktop.

OR

1. Click the **Start** button on the taskbar, choose **All Programs**, then click the **Adobe Illustrator** icon.

   Note: By default a **Welcome Screen** opens when Illustrator launches. You can turn off this option by checking the **Don't Show Again** box.

2. Click the **Close** button on the Welcome Screen.

Click the Adobe Illustrator icon in the menu dock.

launch Adobe Illustrator (mac)

1. Position your cursor toward the bottom of the Mac desktop.

2. Click the **Adobe Illustrator** icon in the menu dock.

   Note: By default a **Welcome Screen** opens when Illustrator launches. You can turn off this option by checking the **Don't Show Again** box.

3. Click the **Close** button on the Welcome Screen

323 - 5̶7̶6̶ 215 - 6243
(323) - 365-8882 - Cynthia
(559) 708 - 7533 - Josephine
- Shirly H.

626-636-6576 Shirly Shi

open croqui/ save as croqui

A croqui is defined as a rough sketch or, in this case, a fashion model. The Electric Fashion CD comes with a standard women's croqui. You will use the women's croqui with the lessons in this book to create precise fashion flats. Be certain to save (Save As) the file to the desktop of your computer.

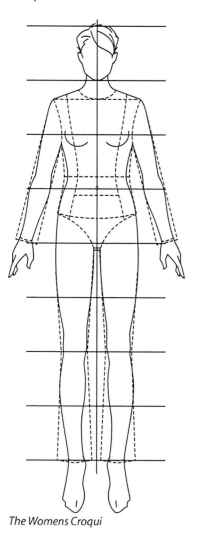

The Womens Croqui

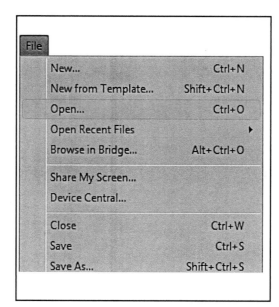

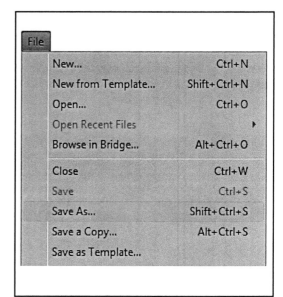

open croqui

1. Place the **electric fashion CD-rom** in the disk drive.

2. Choose **File > Open** from the Illustrator menu bar.

3. In the Open dialog box navigate to **My Computer**, then double-click the disk drive labeled **electric fashion CD**.

4. Double-click to open the **Croqui** folder.

5. Double-click to open the **WomanCroqui** file.

   Note: You can also drag the WomanCroqui file from the Electric Fashion CD to your computers desktop and then double-click to open it.

save as croqui

1. Choose **File > Save As**.

2. In the Save As dialog box navigate to the **Desktop** (click the Desktop icon at left).

3. In the Save As dialog box (**File name:**) type a name for your croqui.

4. Click **Save**.

minimize/ maximize screen

An Illustrator window appears as a standard size on your screen and sometimes smaller than your monitor's screen. To take advantage of a large monitor you will want to maximize the Illustrator screen.

Minimize button
Minimizes the Illustrator document down to the Window's taskbar.

Maximize button
Maximizes the size of the Illustrator document to fit within your entire screen.

Restore Down button
Restores the size of the Illustrator document to its default screen size.

Close button
Closes the Illustrator document and exits the program.

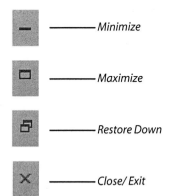

— Minimize

— Maximize

— Restore Down

— Close/ Exit

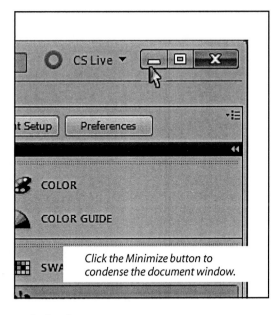

Click the Minimize button to condense the document window.

minimize screen

1. Click the **Minimize** button in the Illustrator Application Control menu to condense it down to the taskbar.

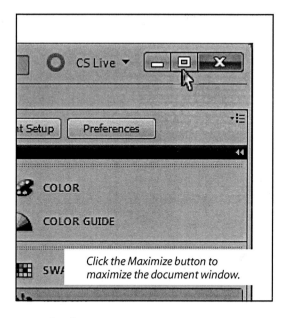

Click the Maximize button to maximize the document window.

maximize screen

1. Click the **Adobe Illustrator** icon in the Window's taskbar to restore the size of the screen.

2. Click the **Maximize** button in the Illustrator Application Control menu to maximize the screen.

> **QUICK TIP!**
> Use the "Change screen mode" option in the Tools panel to choose various screen views.

show panel/ dock panel

Use the Window menu to show commonly used panels. Panels that are floating within the workspace, can be dragged to the left side of the screen to dock them. See page 8 for detailed information on Panels.

Here is a list of commonly used panels:

Align
Brushes
Layers
Navigator
Pathfinder
Stroke
Swatches

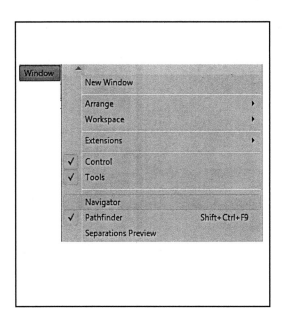

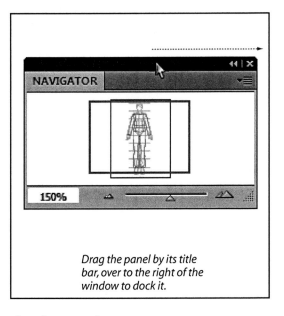

Drag the panel by its title bar, over to the right of the window to dock it.

show panel

1. Choose **Window > Navigator**.

   Note: The panel will either show at the left side dock or it will open as a floating panel. If it opens as a floating panel drag the panel by its title bar, over to the right side of the window to dock it.

dock panel

1. Drag the **Navigator** panel by its title bar, over to the left side of the window until it is docked.

show/ hide views

Illustrator's View menu offers various viewing options. You may choose to show or hide the rulers, show or hide the transparency grid or show or hide the Smart Guides. See page 13 for detailed information on the View menu.

Here is a list of common viewing options found in the View menu:

Show Edges
Show Bounding Box
Show Page Tiling
Show Rulers
Show Transparency Grid
Show Smart Guides

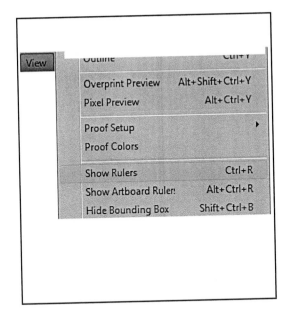

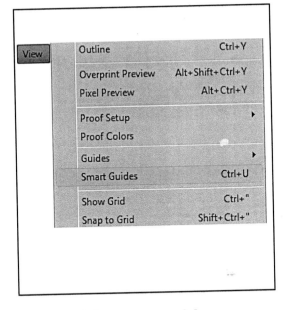

show/ hide rulers

1. Choose **View > Show Rulers**.

   Note: If the menu reads "Hide Rulers" then the rulers are already showing.

2. Choose **View > Hide Rulers** to hide the rulers.

show/ hide smart guides

1. Choose **View > Smart Guides**.

   Note: If the menu shows a check mark before Smart Guides then the smart guides are already showing.

2. Choose **View > Smart Guides** to take away the check and hide the smart guides.

QUICK TIP!
The Tab key on your keyboard will Hide/Show all panels including the Tools panel and the Control Panel.

---

create new layer/ lock layer

Ideally you will want the croqui to remain on a locked layer while the fashion flats are drawn on a second, top layer. Doing this will allow you to draw the fashion flats on the top layer while using the croqui as a guide on the bottom layer.

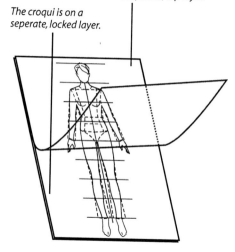

Fashion flats are created on a seperate, top layer.

The croqui is on a seperate, locked layer.

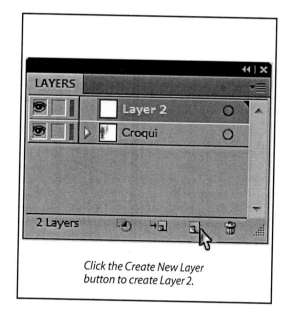

Click the Create New Layer button to create Layer 2.

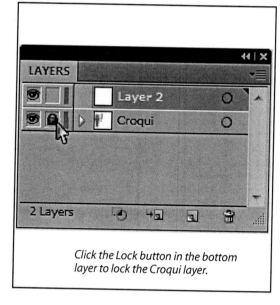

Click the Lock button in the bottom layer to lock the Croqui layer.

create new layer

1. Click the **Layer** panel icon.

    Note: If the Layer panel is not showing choose it from the Window menu.

2. Click the **Create New Layer** button.

lock layer

1. Click the **Lock** box in Croqui layer.

2. Click the new layer (**Layer 2**) to make it active (highlighted).

QUICK TIP!
Having the croqui on a seperate layer is also convienent if you need to hide/show the croqui at any time. Click the eye next to the croqui layer to hide the croqui.

---

zoom in/ scroll page

There are various ways to zoom in/ zoom out and scroll the artboard within a workspace. You will decide which way is most convenient for you. To zoom the artboard you can use keyboard shortcuts or use the Zoom tool or the View menu commands for zooming the page. To scroll the artboard use either the Hand tool or the Navigator panel's Preview Area box. See page 14 for detailed information on Zoom In/ Zoom Out.

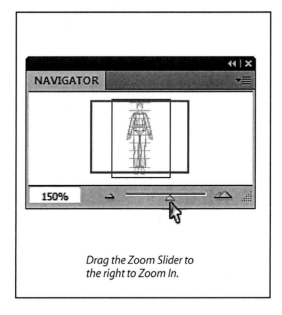

Drag the Zoom Slider to
the right to Zoom In.

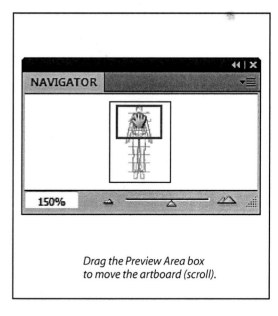

Drag the Preview Area box
to move the artboard (scroll).

QUICK TIP!
Use the Hand tool to move the artboard within the workspace (scroll page). The Spacebar is a keyboard shortcut for the hand tool.

---

zoom in

1. Click the **Navigator** panel.

   Note: If the Navigator panel is not showing choose it from the Window menu.

2. Drag the **Zoom Slider** to the right to zoom in or to the left to zoom out.

scroll page

1. Point to the **Preview Area** within the Navigator panel.

2. Drag the Preview Area (red box) to move the artboard within the workspace (scroll).

scale stroke and effects *(uncheck)*

Often when you are drawing small details such as buttons or zipper pulls you will want to draw them oversized then scale them to fit your fashion flat. Also, when creating various shapes etc., you may need to scale their size either larger or smaller. It is for this reason that you need to ensure that the Scale Stroke & Effect Option is unchecked. Doing this will keep the stroke weight the same no matter how much you scale the object. You can more conveniently set the stroke weight using the Stroke panel or the Control panel.

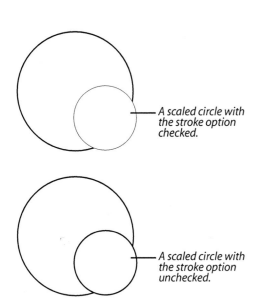

A scaled circle with the stroke option checked.

A scaled circle with the stroke option unchecked.

Double-click the Scale tool to open the scale dialog box.

open scale dialog box

1. Double-click the **Scale** tool in the Tools panel.

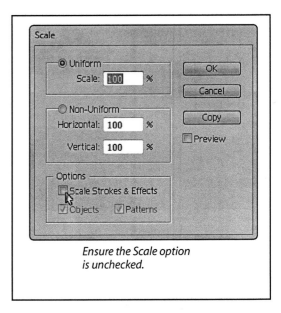

Ensure the Scale option is unchecked.

scale stroke and effects *(uncheck)*

2. Uncheck the **Scale Stroke & Effects** option in the Scale dialog box.

3. Click **OK**.

default fill and stroke

As discussed in chapter two the ideal way to draw fashion flats is by creating closed shapes that overlap each other (bodice, sleeves, neckline). To ensure that your bodice shape (the first part of the fashion flat) has a white fill and a black stroke weight of 1.0 pt (point), click the Default Fill and Stroke option in the Tools panel before drawing the initial bodice shape.

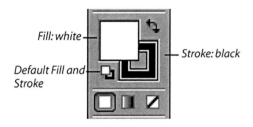

Fill: white

Default Fill and Stroke

Stroke: black

Remember, all style lines (seams), topstitch and drape lines must have a none fill. To set the fill to none before creating a style or drape line, first deselect all objects then click the None option in the Tools panel. Make certain the Fill option in the Tools panel is on top. To change the stroke color click the Stroke box to bring it forward then choose the desired color from the Swatches panel. You can also use the Control panel to change the Stroke and Fill color.

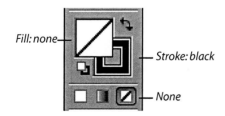

Fill: none

Stroke: black

None

fill and stroke

Keep an eye on the Fill and Stroke box in the Tools panel as you are drawing and selecting objects. This option will display the object's stroke/ fill attributes as it is being drawn or when it is selected.

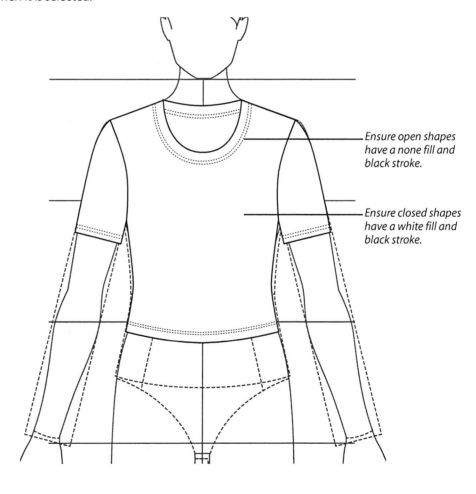

Ensure open shapes have a none fill and black stroke.

Ensure closed shapes have a white fill and black stroke.

chapter five

draw a basic t-shirt
beginner level lesson

When drawing a fashion flat the first task is to create the half bodice shape. This will ensure that both sides are balanced when you reflect the half shape. This technique can be compared to creating a pattern block as in the process of patternmaking.

Using the croqui as your guide, the setting of anchors allow you to create a general shape of the fashion flat choosing from fitted, unfitted, long or cropped shapes.

In this chapter you will create a basic t-shirt with a scoop neckline, short sleeves and topstitch detail. You'll then use a copy of the t-shirt to create the back view, group the completed front and back views and then apply simple color.

create t-shirt shape

Before you begin this tutorial make certain that you have setup your workspace as demonstrated in chapter four (get setup).

Here is a review of how you should setup your workspace:

Launch Adobe Illustrator

Open Croqui/ Save As Croqui

Maximize Screen

Show Panels/ Dock Panels

Show/ Hide Views

Create A New Layer/ Lock Layer 1

Zoom In/ Scroll Page

Scale Stroke & Effects (unchecked)

Default Fill and Stroke

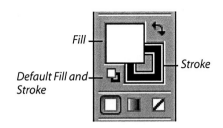

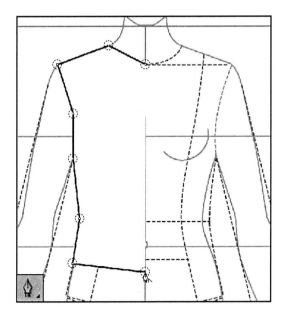

set anchor points

1. Select the **Pen** tool.

2. Click at the **CF neckline** to set the 1st anchor.

   Note: Use the croqui markings as a guide.

3. Click at the **shoulder neckline**.

4. Click at the **shoulder armhole**.

5. Click at the **middle armhole**.

6. Click at the **armpit**.

7. Click at the **waistline**.

8. Click at the **side hip**.

9. Click at the **CF hip**.

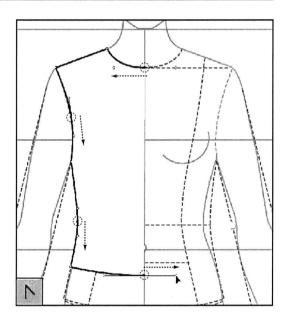

convert anchor points

1. Select the **Convert Anchor Point** tool.

2. Point to the **CF neckline anchor** then drag it horizontally to the left to create a smooth path.

   Note: Be certain not to drag direction points out too far and to keep the direction lines level to ensure smooth shapes.

3. Drag the **middle armhole anchor** downward.

4. Drag the **waistline anchor** downward.

5. Drag the **CF hip anchor** horizontally to the right to create a smooth path.

6. Select the **Selection** tool to cancel the Convert Anchor Point tool.

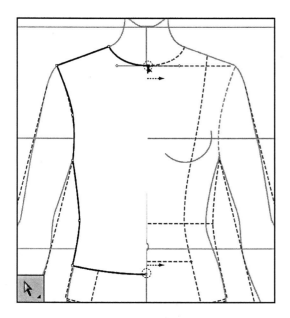

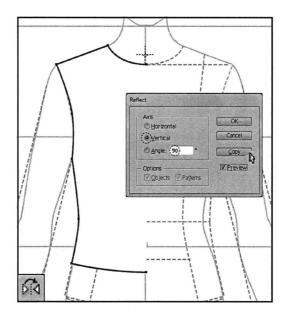

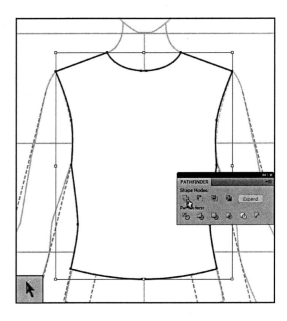

reshape

1. Select the **Direct Selection** tool.

2. Move the **CF neckline anchor** just to the right of the croqui's center front.

3. Move the **CF hip anchor** just to the right of the croqui's center front.

   Note: You want this half t-shirt to cover the croqui's center front so that when it is copy-reflected the two sides overlap.

4. Use the **Direct Selection** tool to move anchors and/or direction handles to reshape and perfect the half t-shirt.

reflect

1. Select the half t-shirt with the **Selection** tool.

2. Select the **Reflect** tool then hold the **Alt key** on the keyboard.

3. **Alt+click** at the CF guideline of the croqui to set the reflect origin.
   (Mac) Option + click

4. In the **Reflect** dialog box click the **Vertical** button and type **(90)** for the **Angle**. Click **Copy**.

5. Select the **Selection tool** to cancel the Reflect tool.

6. Nudge the half t-shirt into place using the **Right/ Left Arrow key** on the keyboard.

   Note: Be certain the shapes are overlapping.

unite

1. Select the **Selection** tool then hold the **Shift key** on your keyboard.

2. **Shift+click** the left half t-shirt then the right half t-shirt to select both halves.

3. In the **Pathfinder** panel click the **Unite** button.

   Note: If using earlier versions of Adobe Illustrator (CS2 or CS3) Be certain to click the Expand button.

4. Select the **Selection** tool then click outside the t-shirt to deselect.

   Note: The two shapes are combined to create one closed shape.

create neckline

Now that you have the t-shirt bodice drawn you will create a neckline shape, a closed shape with a white fill, that is arranged on top of the t-shirt bodice. To create a basic neckline you can use the t-shirt bodice as a guide to set anchors in the general shape of the neckline.

To create a more complex neckline shape you can create half the neckline, using the croqui center front as a guide, and then reflect the half neckline, similar to how the bodice was created. You'll then overlap the two sides and use the Pathfinder panel to combine the two shapes.

Create the half neckline shape.

Reflect the neckline shape then Add to shape area/ Expand.

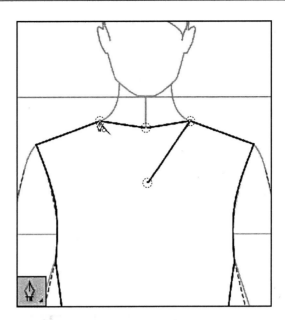

set anchor points

1. Select the **Pen** tool.

2. Click at the **left shoulder neckline** to set the 1st anchor.

3. Click at the **center back neckline** just above the bodice neckline.

4. Click at the **right shoulder neckline**.

5. Click at the **center front neckline** just inside the bodice.

6. Point to the **left shoulder neckline** (the first anchor set).

    Note: Look for the "Closing a path" symbol.

7. Click inside the **left shoulder neckline** to close the shape.

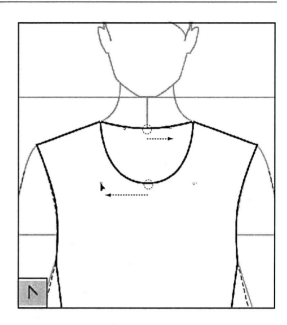

convert anchor points

1. Select the **Convert Anchor Point** tool.

2. Drag the **CB neckline anchor** horizontally to the right to create a smooth path.

    Note: Be certain not to drag direction points out too far and to keep the direction lines horizontally level to ensure a smooth neckline.

3. Drag the **CF neckline anchor** horizontally to the left.

4. Use the **Direct Selection** tool to move anchors and/or direction lines to reshape and perfect the neckline shape.

create sleeve

The sleeve you will create here is a basic set in sleeve. To create the sleeve you will copy the armhole path (Edit > Copy), paste it directly on top of the bodice armhole (Edit > Paste In Front) then use the Pen tool to pick up the path and close the shape.

To add details to your sleeve such as a style line or cuff first create the sleeve shape, deselect it then use the Pen tool or the shape tools to create the detail.

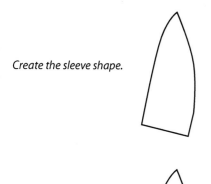

Create the sleeve shape.

Create the sleeve detail on top of the sleeve shape.

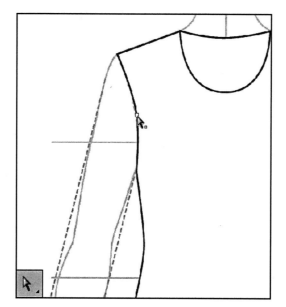

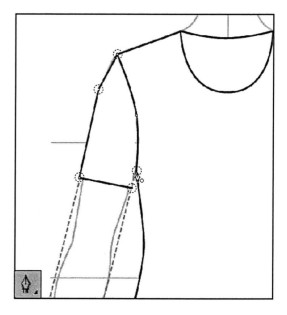

copy/ paste path

1. Select the **Direct Selection** tool.

2. Point to the **middle armhole anchor**.

3. Click to select the **middle armhole anchor**.

   Note: The anchor will highlight red to indicate that it is selected.

4. Choose **Edit > Copy** then **Edit > Paste in Front**.

   Note: This will paste the armhole path directly on top of the bodice armhole.

set anchor points

1. Select the **Pen** tool.

2. Point to the **shoulder armhole anchor**.

   Note: Look for the "Connecting a path" symbol.

3. Click inside the **shoulder armhole anchor** to pick up the path.

4. Click at the **shoulder ball** to set an anchor.

5. Click at the **outside arm** then the **inside arm** (sleeve hem).

6. Point to the **armpit anchor**.

   Note: Look for the "Closing a path" symbol.

7. Click inside the **armpit anchor** to close the shape.

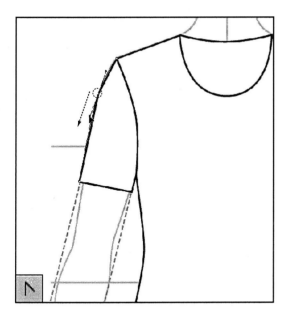

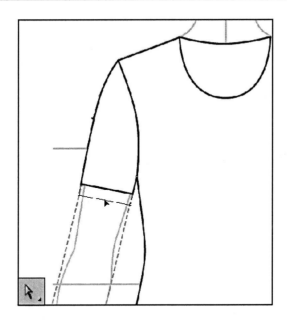

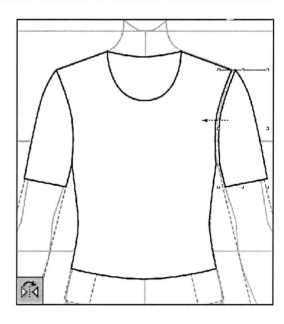

convert anchor points

1. Select the **Convert Anchor Point** tool.

2. Drag the **shoulder ball anchor** downward to create a smooth anchor.

reshape

1. Select the **Direct Selection** tool.

2. Click to select the **shoulder ball anchor**.

3. Drag the anchor or its direction points to create more shape and roundness to the sleeve.

4. Click to select the sleeve hem (path) between its two anchors.

   Note: Be certain to click directly on the path. The anchors will highlight (white) when the path is selected.

5. Drag the path downward to create more length to the sleeve.

reflect

1. Select the sleeve with the **Selection** tool.

2. Select the **Reflect** tool then hold the **Alt key** on the keyboard.

3. **Alt+click** at the CF guideline of the croqui to set the reflect origin. *(Mac) Option+click*

4. In the **Reflect** dialog box click the **Vertical** button and type **(90)** for the **Angle**. Click **Copy**.

5. Select the **Selection tool** to cancel the Reflect tool.

6. Nudge the sleeve into place using the **Right/ Left Arrow key** on your keyboard.

QUICK TIP!
A convenient way to undo/redo is by using the keyboard shortcut. Press **Ctrl+Z** on your keyboard to Undo an action. *(Mac) Command+Z*

---

add topstitch

The ideal way to add topstitch to a fashion flat is by copying a path (Edit > Copy), pasting it onto the page (Edit > Paste) then using the Stroke panel to set topstitch attributes (Weight and Dashed Line). Style lines, drape lines and as well topstitching are most commonly open paths and should be created with a none fill.

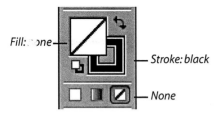

Fill: none —

— *Stroke: black*

— *None*

Set the path's stroke attributes in the Stroke Panel. You can access the Stroke panel (Stroke:) through the Control Panel or use the Stroke panel at the right of the screen.

Show/ Hide Options.

Set the Stroke Weight.

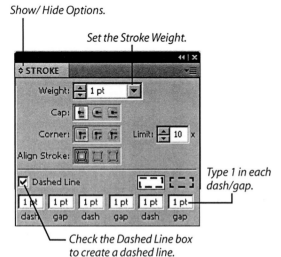

Type 1 in each dash/gap.

Check the Dashed Line box to create a dashed line.

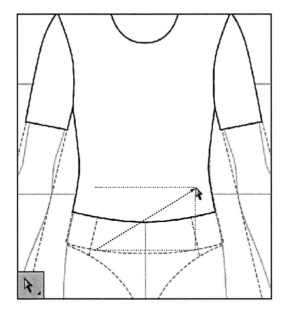

copy/ paste path

1. Select the **Direct Selection** tool.

2. Drag a selection marquee from beneath the t-shirt then up and over the **CF hip anchor(s)** to select them.

   Note: The anchor(s) will highlight red to indicate that they are selected.

3. Choose **Edit > Copy** then **Edit > Paste**.

   Note: This will paste the hemline path onto the page.

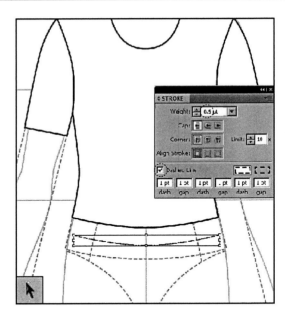

none fill/ dashed line

1. WIth the hemline path selected click **None** in the **Tools** panel.

2. In the **Stroke** panel click the **Weight** downward arrow to choose a stroke weight of **(0.5pt).**

3. Click the **Dashed Line** button.

4. Type **(1pt)** in each **dash/ gap** field.

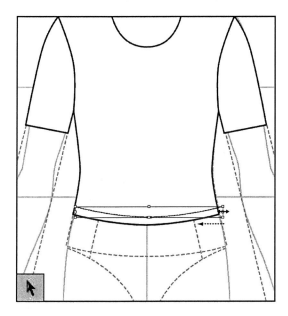

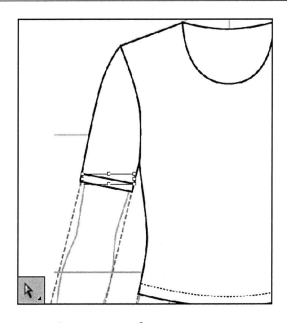

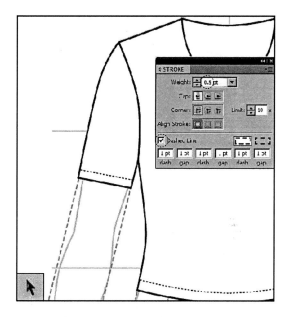

reshape

1. Select the hemline path with the **Selection** tool.

   Note: The path does not have a fill so you must click/ drag it by its stroke (outline).

2. Move the path into position at the hem of the t-shirt bodice.

3. Drag the path's side bounding box handles to fit it onto the t-shirt hemline.

QUICK TIP!
For two rows of topstitch duplicate the topstitch path. To duplicate a path select it with the Selection tool then choose Edit > Copy then Edit > Paste.

---

copy/ paste path

1. Select the **Direct Selection** tool.

2. Point to the path at the sleeve hem.

3. Click to select the sleeve hem.

   Note: The sleeve anchors will highlight (white) to indicate the path is selected.

4. Choose **Edit > Copy** then **Edit > Paste**.

   Note: This will paste the sleeve hemline path onto the page.

none fill/ dashed line

1. With the sleeve path selected click **None** in the **Tools** panel.

2. In the **Stroke** panel click the **Weight** downward arrow to choose a stroke weight of **(0.5pt).**

3. Click the **Dashed Line** button.

4. Type **(1pt)** in each **dash/ gap** field.

5. Select the copied path with the **Selection** tool.

   Note: The path does not have a fill so you must click/ drag it by its stroke (outline).

6. Move the path into position at the hem of the sleeve.

7. Use the **Reflect tool** to reflect the topstitch.

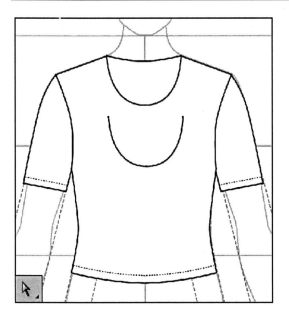
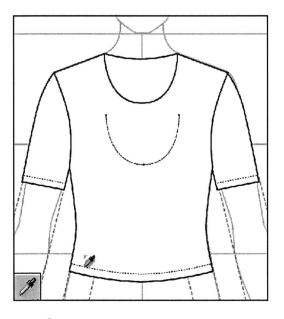
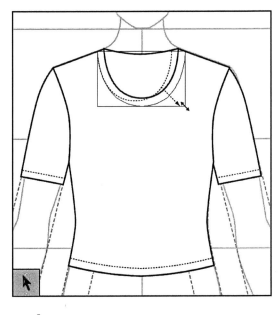

copy/ paste path

1. Select the **Direct Selection** tool.

2. Point to the **CF neckline anchor**.

3. Click to select the **CF neckline anchor**.

   Note: The anchor will highlight blue to indicate that it is selected.

4. Choose **Edit > Copy** then **Edit > Paste**.

   Note: You may also choose to have topstitch on the inside of the neckline. To do this select the CB neckline anchor and choose Edit > Copy then Edit > Paste.

eyedropper (none fill/ dashed line)

1. With the neckline path selected click to select the **Eyedropper** tool.

   Note: The Eyedropper serves as a quick way to get a none fill/ dashed line for the topstitch.

2. Click on the topstitch at the hem of the t-shirt to sample its topstitch attributes.

reshape

1. Select the neckline path with the **Selection** tool.

   Note: The path does not have a fill so you must click/ drag it by its stroke (outline).

2. Move the path into position at the left side of t-shirt neckline.

3. Drag the path's corner bounding box handles to scale it.

create back view

Before you continue drawing be certain that you have saved what you have done thus far. As you started this chapter you should have saved this file onto your computers desktop (File > Save As). Choose File > Save to save an updated version of your document.

The back view of a fashion flat often times has the same shape as a front view only with varying (back view) details. Here you will create the back view of your basic t-shirt using a copy of the front view.

The basic t-shirt back view is a copy of the front view with changes made to the neckline.

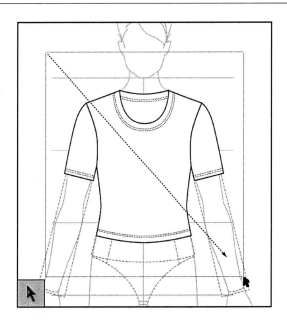

copy/ paste t-shirt

1. Select the **Selection** tool.

2. Drag from outside the t-shirt then over and down to create a selection marquee.

3. Choose **Edit > Copy** then **Edit > Paste**.

4. Move the copied t-shirt so that it is right next to the original.

   Note: Do not overlap the copied t-shirt until you have completed both views and grouped them seperately.

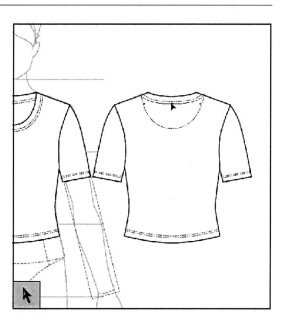

delete details

1. Select the neckline shape on the copied t-shirt with the **Selection** tool.

2. Press the **Delete key** on your keyboard.

reshape

1. Select the **Direct Selection** tool.

2. Move the **CB neckline anchor** (bodice) upward so as to appear as the back view.

3. Move the topstitch anchors upward so as to appear as the back view topstitch.

group details/ group items

Now that you have completed the t-shirt front and back views you can group the seperate details and then group the complete front view and complete back view. This will allow you to freely move fashion flats as one object and overlap them as in a presentation.

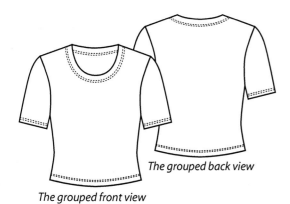

The grouped back view

The grouped front view

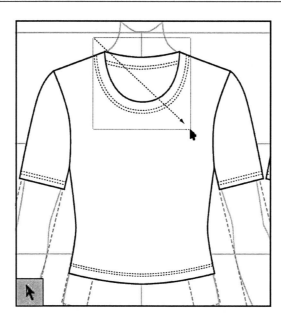

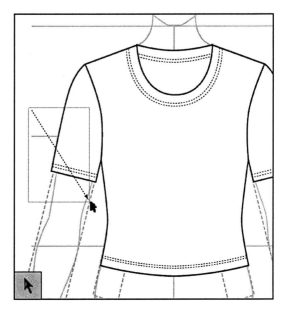

group front neckline

1. Select the **Selection** tool.

2. Drag from outside the t-shirt then over and down to select the neckline parts.

   Note: The t-shirt bodice is included in the selection, you will use the Shift key to release it from the selection.

3. Hold the **Shift key** on your keyboard then **Shift+click** the t-shirt bodice to release it from the selection.

4. Choose **Object > Group** to group the neckline parts.

group front sleeves

1. With the **Selection** tool drag from outside the t-shirt then over and down to select the left side sleeve parts.

2. Choose **Object > Group.**

3. With the **Selection** tool drag from outside the t-shirt then over and down to select the right sleeve parts.

4. Choose **Object > Group.**

> **QUICK TIP!**
> A convenient way to choose the Group or Ungroup command is by using the onscreen context-sensitive menu. To do this select the desired objects then right-click on the artboard to choose Group. *(Mac) Control-click*

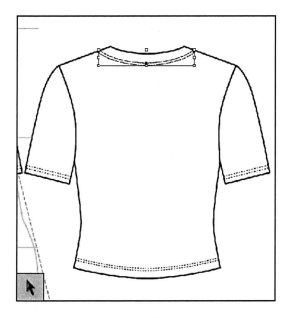

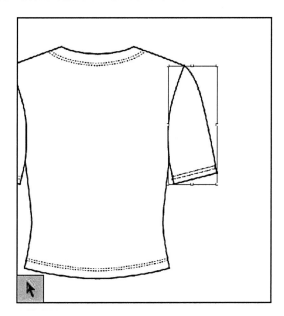

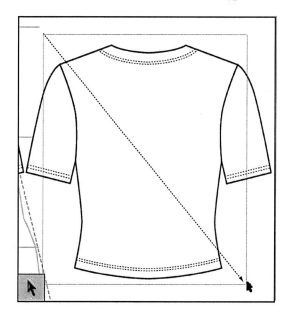

group back neckline

1. With the **Selection** tool drag from outside the t-shirt then down and over to select the back neckline parts.

   Note: The t-shirt shape is included in the selection, you will use the Shift key to release it from the selection.

3. Hold the **Shift key** on your keyboard then **Shift+click** the back t-shirt bodice to release it from the selection.

4. Choose **Object > Group**.

group back sleeves

1. With the **Selection** tool drag from outside the t-shirt then over and down to select the left side sleeve parts.

2. Choose **Object > Group**.

3. With the **Selection** tool drag from outside the t-shirt then over and down to select the right sleeve parts.

4. Choose **Object > Group**.

group entire front

1. With the **Selection** tool drag from outside the t-shirt then over and down to select the t-shirt front view.

2. Choose **Object > Group**.

group entire back

1. With the **Selection** tool drag from outside the t-shirt then over and down to select the t-shirt back view.

2. Choose **Object > Group**.

add color

When applying color to your fashion flats keep in mind that you only want to add color to its closed shapes (bodice, sleeve, neckline). If you add color to open paths (topstitch, style lines) you will get unusual results that make parts of your sketch appear covered up.

Color added to topstitch appears incorrect.

Color added to the bodice, sleeve and neckline only.

When applying color to the fashion flat be certain that the Fill box in the Tools panel is on top. This ensures that you are applying color to the fill and not the stroke.

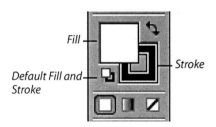

Fill

Default Fill and Stroke

Stroke

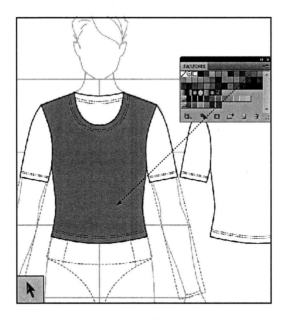

drag and drop color

1. From the **Swatches** panel drag the desired color over to a part of the t-shirt.

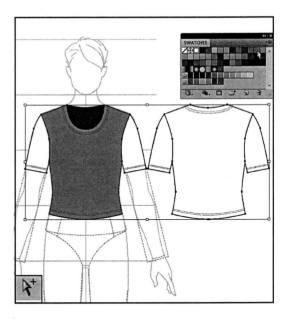

group select and color

1. Select the **Group Selection** tool then hold the **Shift key** on your keyboard.

2. **Shift+click** the parts of the t-shirt that will have color added to them (closed shapes).

3. In the **Swatches** panel click the desired color.

QUICK TIP!
Use the Color panel to mix colors (Cyan, Magenta, Yellow and Black) or pick colors from the Color panel's Spectrum.

---

explore t-shirt design

Congradulations designer, you have completed
a basic t-shirt, front and back view. From here
you can explore basic t-shirt designs with great
ease. Begin by making a copy of the original
t-shirt, ungroup it, reshape parts of it (sleeve,
neckline), remove details then add new details,
topstitch and/or style lines.

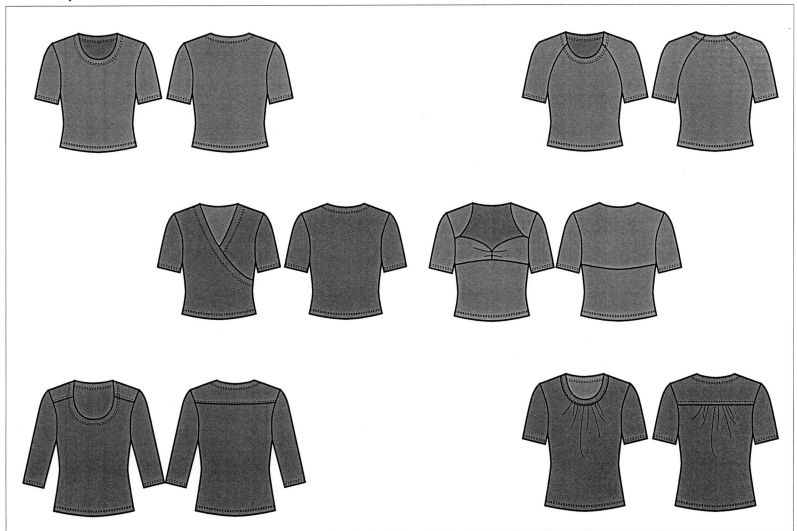

chapter six

draw a tailored shirt
beginner level lesson

A tailored shirt is defined as an upper body garment with a collar, sleeves, and a button front placket. A tailored shirt can also be called a dress shirt or in women's clothing a shirt is also reffered to as a blouse.

The approach to drawing the tailored shirt in this chapter is similar to drawing the basic t-shirt, as demonstrated in chapter five. Again you will use the croqui as a guide, setting anchors to create the general shape of the tailored shirt. Once the tailored shirt bodice is complete with its style lines and topstitch you'll then create the collar and its details, then the sleeves and its details. To complete the tailored shirt you will use a copy of its front view to create the back view, group the completed front and back views and then apply color.

create shirt shape

Before you begin I must remind you to always setup your workspace as demonstrated in chapter four (get setup). Doing this will ensure that your fashion flats are precise, proportioned and easy to edit and that your workspace is easy to navigate.

Setup your workspace before you begin:

Launch Adobe Illustrator

Open Croqui/ Save as Croqui

Maximize Screen

Show Panels/ Dock Panels

Show/ Hide Views

Create A New Layer/ Lock Layer 1

Zoom In/ Scroll Page

Scale Stroke & Effects (unchecked)

Default Fill and Stroke

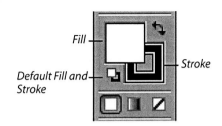

Fill

Stroke

Default Fill and Stroke

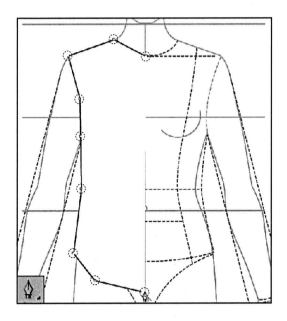

set anchor points

1. Select the **Pen** tool.

2. Click at the **CF neckline** to set the 1st anchor.

3. Click at the **shoulder neckline**.

4. Click at the **shoulder armhole**.

5. Click at the **middle armhole**.

6. Click at the **armpit**.

7. Click at the **waistline**.

8. Click at the **side hip**.

8. Click at the **princess hip**.

9. Click at the **CF hip**.

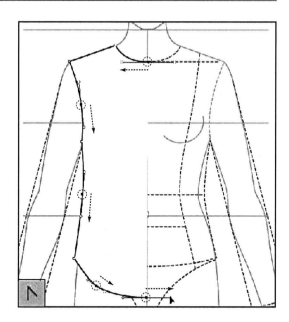

convert anchor points

1. Select the **Convert Anchor Point** tool.

2. Point to the **CF neckline anchor** then drag it horizontally to the left to create a smooth path.

3. Drag the **middle armhole anchor** downward.

4. Drag the **waistline anchor** downward.

5. Drag the **princess hip anchor** diagonally downward toward the CF hip.

6. Drag the **CF hip anchor** horizontally to the right to create a smooth path.

7. Select the **Selection** tool to cancel the Convert Anchor Point tool.

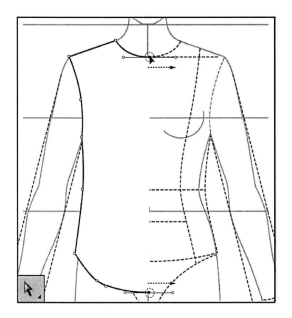

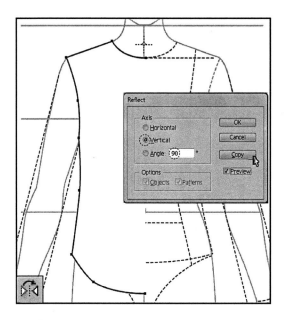

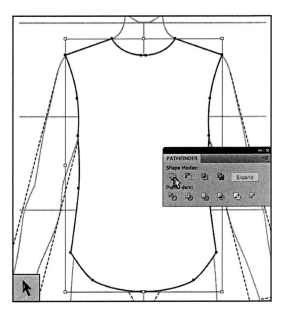

reshape

1. Select the **Direct Selection** tool.

2. Move the **CF neckline anchor** just to the right of the croqui's center front.

3. Move the **CF hip anchor** just to the right of the croqui's center front.

   Note: You want this half tailored shirt to cover the croqui's center front so that when it is copy-reflected the two sides overlap.

4. Use the **Direct Selection** tool to move anchors and/or direction handles to reshape and perfect the half tailored shirt.

reflect

1. Select the half tailored shirt with the **Selection** tool.

2. Select the **Reflect** tool then hold the **Alt key** on the keyboard.

3. **Alt+click** at the CF guideline of the croqui to set the reflect origin.
   (Mac) Option+click

4. In the **Reflect** dialog box click the **Vertical** button and type **(90)** for the **Angle**. Click **Copy**.

5. Select the **Selection tool** to cancel the Reflect tool.

6. Nudge the half shirt into place using the **Right/ Left Arrow key** on the keyboard.

   Note: Be certain the shapes are overlapping.

unite

1. Select the **Selection** tool then hold the **Shift key** on your keyboard.

2. **Shift+click** the left half shirt then the right half shirt to select both halves.

3. In the **Pathfinder** panel click the **Unite** button.

4. Select the **Selection** tool then click outside the shirt to deselect.

QUICK TIP!
The Shape Builder tool in the Tools panel is an easy way to unite two or more shapes. Select two or more shapes then drag the Shape Builder tool across them to combine and create one shape.

---

add topstitch and style lines

To ensure the hemline topstitch is balanced with the shape of the shirt you will (Copy/ Paste) the path then set its topstitch attributes using the Stroke panel.

By default the Line Segment tool has a none fill making it the most convenient of the drawing tools for creating style lines. Remember, topstitch and style lines should always be created with a none fill.

To create a princess seam use the Line Segment tool to draw a straight line from the shirt armhole to its hemline then use the Convert Anchor Point tool to convert it into a smooth path.

Style lines created with the Line Segment and Convert Anchor Point.

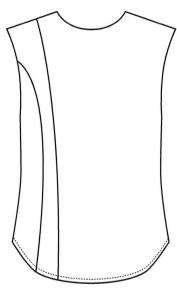

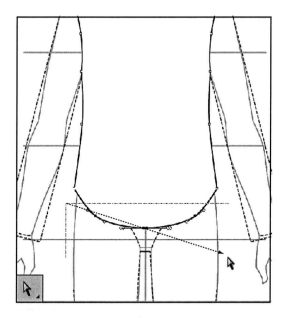

copy/ paste path

1. Select the **Direct Selection** tool.

2. Drag a selection marquee from beneath the shirt then up and over the **hemlines anchors** to select them.

   Note: Only select the hemline anchors not the side hip anchors.

3. Choose **Edit > Copy** then **Edit > Paste**.

reshape

1. Select the copied path with the **Selection** tool.

2. Move the path into position at the hem of the tailored shirt.

3. Drag the path's side bounding box handles to fit it onto the tailored shirt hemline.

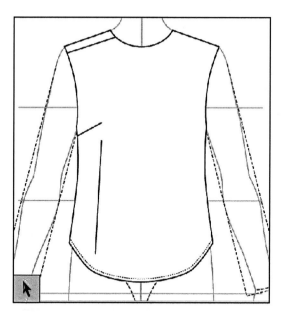

none fill/ dashed line

1. With the hemline path selected click **None** in the **Tools** panel.

2. In the **Stroke** panel click the **Weight** downward arrow to choose a stroke weight of **(0.5pt).**

3. Click the **Dashed Line** button.

4. Type **(1pt)** in each **dash/ gap** field.

create style lines

1. Deselect then click the **Default Fill and Stroke** option in the Tools panel.

2. Select the **Line Segment** tool then drag to create style lines on the left side of the tailored shirt.

3. Use the **Reflect tool** to reflect the style lines.

create collar

Collars and lapels are created with four simple shapes, the left collar shape, the right collar shape, the back collar shape and the inside lining shape. All four shapes have a white fill and are stacked with the left and right collar on top and the back and lining shapes sent to back. The back collar shape has dual use as it can be copied and used on the back view of the tailored shirt.

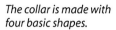

The collar is made with four basic shapes.

The back collar shape will also be used on the back view of the shirt.

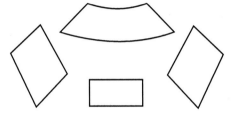

The shapes have a white fill and are stacked.

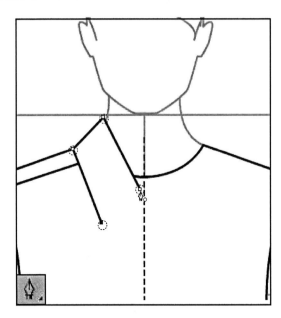

set anchor points

1. Click the **Default Fill and Stroke** option in the Tools panel.

2. Select the **Pen** tool.

3. Click slighty to the left of the **CF neckline** to set the 1st anchor.

4. Click at the **raised neckline**.

5. Click at the **shoulder**.

6. Click at the **collar point**.

7. Point to the **CF neckline** (the 1st anchor).

   Note: Look for the "Closing a path" symbol.

8. Click inside the **CF neckline anchor** to close the shape.

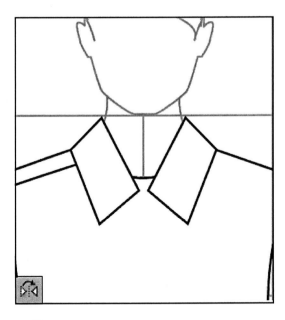

reflect

1. Select the collar shape with the **Selection** tool.

2. Select the **Reflect** tool then hold the **Alt key** on the keyboard.

3. **Alt+click** at the CF guideline of the croqui to set the reflect origin.
   (Mac) Option+click

4. In the **Reflect** dialog box click the **Vertical** button and type **(90)** for the **Angle**. Click **Copy**.

5. Select the **Selection tool** to cancel the Reflect tool.

6. Nudge the collar into place using the **Right/ Left Arrow key** on your keyboard.

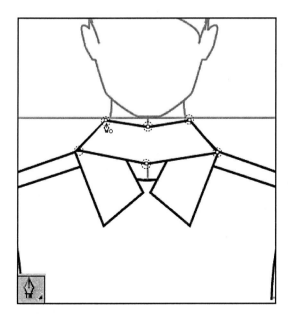

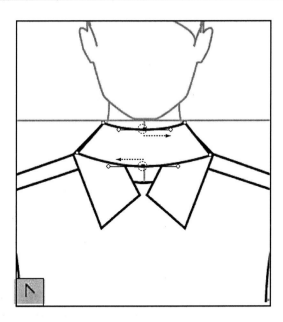

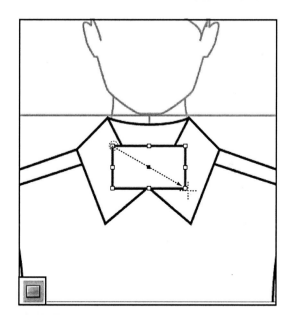

set anchor points

1. Select the **Pen** tool.

2. Click at the left **raised neckline** to set the 1st anchor.

3. Click at the **CB neckline** (upper).

4. Click at the right **raised neckline**.

5. Click at the **right shoulder**.

6. Click at the **CB neckline** (lower).

7. Click a the **left shoulder**.

8. Point to the left **raised neckline**.

   Note: Look for the "Closing a path" symbol.

9. Click inside the **raised neckline anchor** to close the shape.

convert anchor points

1. Select the **Convert Anchor Point** tool.

2. Drag the **CB neckline anchor** (upper) horizontally to the right to create a smooth path.

3. Drag the **CB neckline anchor** (lower) horizontally to the left to create a smooth path.

send to back

1. Select the back collar shape with the **Selection** tool.

2. **Right-click** the mouse to open the context menu. (*Mac*) *Control-click*

3. Choose **Arrange > Send to Back**.

create a rectangle

1. Select the **Rectangle** tool.

2. Drag diagonally to create a small rectangle on top of the collar opening.

   Note: End the rectangle at the CF collar opening.

send to back

1. **Right-click** the mouse to open the context menu. (*Mac*) *Control-click*

2. Choose **Arrange > Send to Back**.

3. Select the **Selection** tool then click outside the shirt to deselect.

add collar topstitch

Now you will add topstitch to the collar. Here you will set the topstitch stroke attributes before you use the Pen tool to set anchor points. Make certain that no objects are selected before setting the topstitch stroke attributes.

Remember to zoom into the collar so that you can see the details clearly.

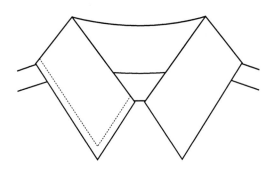

Details are easier to see when zoomed in.

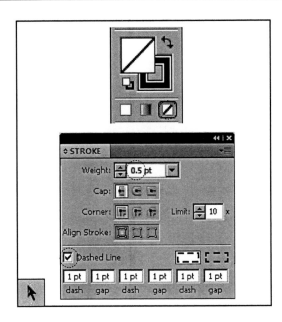

none fill/ dashed line

1. Deselect all objects with the **Selection** tool.

2. Click **None** in the Tools panel to turn off the fill.

3. In the **Stroke** panel click the **Weight** downward arrow to choose a stroke weight of **(0.5pt)**.

4. Click the **Dashed Line** button.

5. Type **(1pt)** in each **dash/ gap** field.

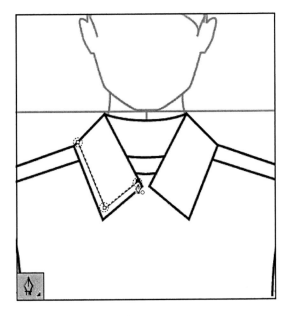

set anchor points

1. Select the **Pen** tool.

2. Click inside the **left collar shoulder** to set the 1st anchor.

3. Click inside the **left collar point**.

4. Click inside the **left CF neckline**.

reflect

1. Select the collar topstitch with the **Selection** tool.

2. Select the **Reflect** tool then **Alt+click** at the CF guideline of the croqui to set the reflect origin. *(Mac) Option+click*

4. In the **Reflect** dialog box click **Copy**.

create placket

The shirt placket is created with three straight lines, one for the collar seam (CF neckline), one for the placket (shirt openning), and one for the topstitch of the placket. Holding the Shift while drawing with the Line Segment tool will constrain the line and create a perfectly straight line.

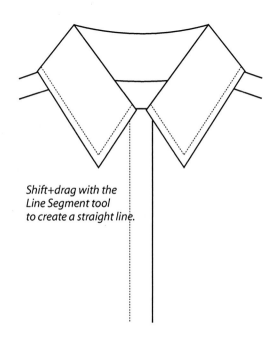

*Shift+drag with the
Line Segment tool
to create a straight line.*

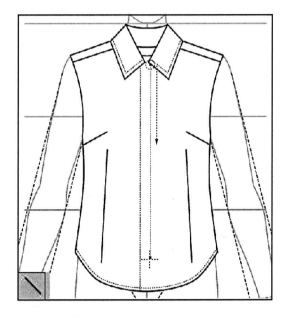

create line

1. Select the **Line Segment** tool.

2. Hold the **Shift key** on your keyboard.

3. **Shift+drag** from the left collar down to the shirt hemline to create a straight topstitch line.

4. **Shift+drag** from the collar (right) down to the shirt hemline to create a straight closure line.

5. In the **Stroke** panel click the **Weight** downward arrow to choose a stroke weight of **(1 pt)**.

6. Click the **Dashed Line** button to turn off the dash option.

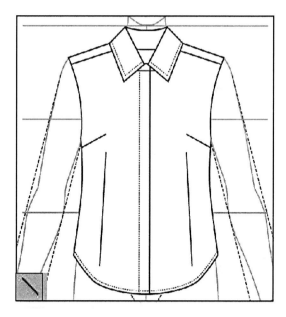

create line

1. With the **Line Segment** tool selected hold the **Shift key** on your keyboard.

2. **Shift+drag** from one side of the collar (inside left) over to the other side of the collar (inside right) to create the collar seam.

3. Select the **Selection** tool then click outside the shirt to deselect.

QUICK TIP!
Did you turn off the Scale Stroke and Effects in the Scale tool? Double click the Scale tool in the Tools panel and uncheck the option.

---

create button

The button and buttonhole is created with the Ellipse tool, the Rectangle tool and the Line Segment tool. To create a perfect Ellipse (circle) you will hold the Shift and Alt keys as you draw the shape. Create the button oversized and away from the shirt, group it and then scale it down to the desired size. The button is a small detail so set the stroke weight to 0.5 pt in the Stroke panel.

The button is made with a few basic shapes.

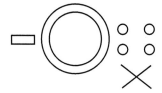

The shapes have a white fill and are stacked.

> **QUICK TIP!**
> Where you have several buttons that are grouped individually, select them all then use the Align panel to align and distribute them.

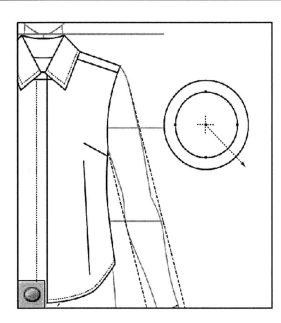

create ellipse

1. Click the **Default Fill and Stroke** option in the Tools panel.

2. Select the **Ellipse** tool.

3. Hold the **Shift key** and **Alt key** on your keyboard.

4. **Shift+Alt+drag** diagonally to create an ellipse. *(Mac) Shift+Option+drag*

5. With the Ellipse tool still selected point to the center of the ellipse you just created.

6. **Shift+Alt+drag** diagonally to create a second ellipse inside the first one.

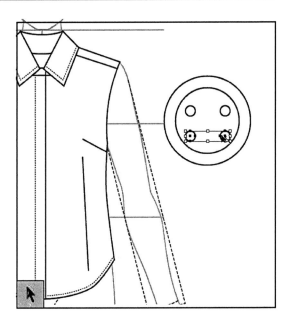

duplicate ellipse

1. With the **Ellipse** tool still selected **Shift+drag** diagonally to create a smaller ellipse for the thread hole.

2. Select the **Selection** tool then hold the **Shift key** and **Alt key** on your keyboard.

3. **Shift+Alt+drag** the thread hole over to the right side of the button to duplicate it. *(Mac) Shift+Option+drag*

4. Continue holding the **Shift key** and **Alt key** then **Shift+Alt+drag** both thread holes down to the bottom of the button to duplicate them.

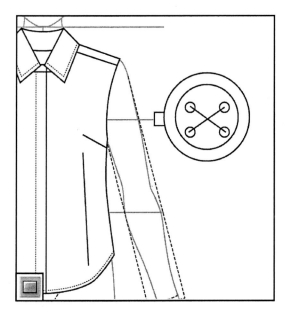

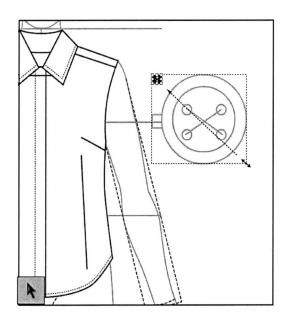

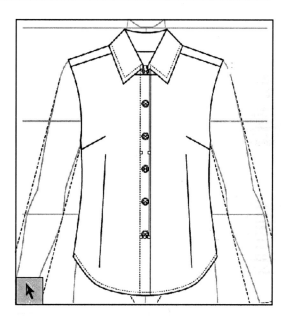

create rectangle

1. Select the **Rectangle** tool.

2. Drag diagonally to create a rectangle for the buttonhole.

send to back

1. Select the rectangle.

2. **Right-click** then choose **Arrange > Send to Back**. *(Mac) Ctrl-click*

create line

1. Select the **Line Segment** tool.

2. Drag diagonally from the top thread hole to the bottom thread hole.

3. Repeat for the opposite thread hole.

group button

1. Select the **Selection** tool.

2. Drag to create a selection marquee around the entire button.

3. **Right-click** the mouse button and choose Group. **(Mac) Ctrl-click**

scale button

1. Hold the **Shift key**.

2. **Shift+drag** the button's corner handle (bounding box) until it is the desired size.

set stroke weight

1. In the **Stroke** panel click the **Weight** downward arrow to choose a stroke weight of **0.5 pt**.

duplicate and step repeat

1. Select the **Selection** tool.

2. Position the button onto the shirt placket.

3. **Shift+Alt+drag** the button down to the second position to move and duplicate it. *(Mac) Shift+Option+click*

4. Press **Ctrl+D** on your keyboard to **Step Repeat**. Repeat Ctrl+D for the number of buttons desired. *(Mac) Cmd+D*

group buttons

1. Select the **Selection** tool.

2. **Shift+click** each button to select them all.

3. Choose **Object > Group**.

create sleeve

Here you will create a long, set in sleeve with a basic cuff and topstitch. To create the initial shape of the sleeve you will copy the armhole path, paste it directly on top of the bodice armhole then use the Pen tool to pick up the path and close the shape. The cuff and its topstitch will be drawn seperately.

To add drape lines to your sleeve you can use the Line Segment tool to draw in simple straight lines then use the Convert Anchor Point tool to make them smooth. Drape lines look best at a stroke weight of 0.5 pt.

Drape lines are created with the Line Segment and Convert Anchor Point.

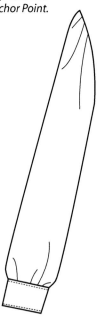

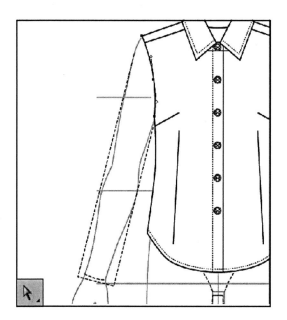

copy/ paste path

1. Select the **Direct Selection** tool.

2. Point to the **middle armhole anchor**.

    Note: The anchor will highlight red to indicate that it is selected.

3. Click to select the **middle armhole anchor**.

4. Choose **Edit > Copy** then **Edit > Paste In Front**.

    Note: This will paste the armhole path directly on top of the bodice armhole.

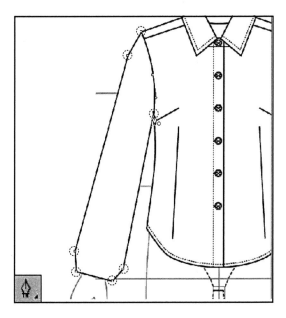

set anchors

1. Select the **Pen** tool.

2. Click inside the **shoulder armhole anchor** to pick up the path.

Note: Look for the "Connecting a path" symbol.

3. Click at the **shoulder ball** to set an anchor.

4. Click at the **upper wrist**.

5. Click a the **outside wrist** then the **inside wrist** (sleeve hem).

6. Click at the **upper wrist**.

7. Click inside the **armpit anchor** to close the shape.

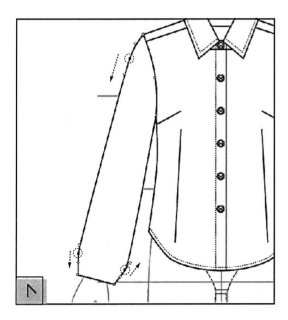

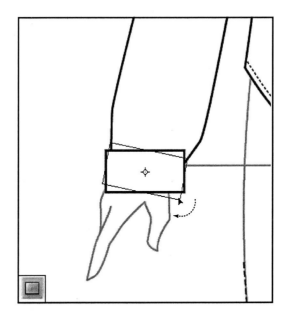

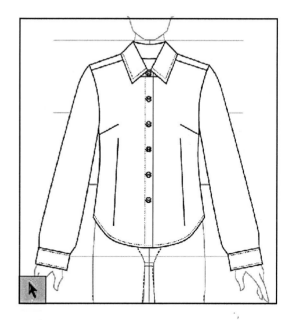

convert anchor points

1. Select the **Convert Anchor Point** tool.

2. Drag the **shoulder ball anchor** down toward the wrist to create a smooth anchor.

3. Drag the **upper wrist anchor** (outside) down toward the wrist to create a smooth anchor.

4. Drag the **upper wrist anchor** (inside) down toward the wrist to create a smooth anchor.

create rectangle

1. Select the **Rectangle** tool.

2. Drag diagonally to create a rectangle.

rotate

1. Select the **Rotate** tool.

2. Drag the rectangle to rotate it.

add topstitch

1. Select **Line Segment** tool.

2. Drag to create a line on top of the cuff.

3. In the **Stroke** panel click the **Weight** downward arrow to choose a stroke weight of **0.5 pt.** Click the **Dashed Line** button to turn on the dash option.

reflect

1. Select the **Selection** tool then hold the **Shift key** on your keyboard.

2. **Shift+click** each part of the sleeve to select all its parts.

3. Select the **Reflect** tool then hold the **Alt key** on your keyboard.

4. **Alt+click** at the CF guideline of the croqui to set the reflect origin. *(Mac) Option+click*

5. In the **Reflect** dialog box click the Vertical button and type (90) for the Angle. Click **Copy**.

6. Nudge the sleeve into place using the **Right/ Left Arrow key** on your keyboard.

create back view

Have you saved your work thus far? Choose File > Save or Save As to save what you have done thus far.

Now to create the back view of your tailored shirt. The back view of the tailored shirt is simply a copy of the front view with varying details. Use various drawing tools to draw the back view style lines, drape and topstitch.

The tailored shirt back view is a copy of the front view with changes made to the neckline, bodice and sleeve.

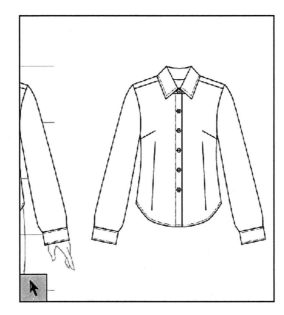

copy/ paste tailored shirt

1. Select the **Selection** tool.

2. Drag from outside the tailored shirt then over and down to create a selection marquee.

3. Choose **Edit > Copy** then **Edit > Paste**.

4. Move the copied shirt so that it is next to the original.

   Note: Do not overlap the tailored shirt views until you have completed them and grouped them seperately.

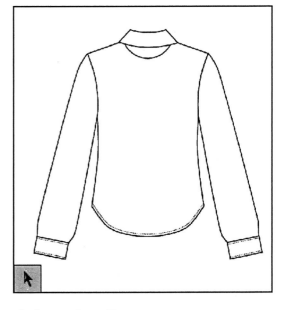

delete details

1. Select and delete the collar shapes and topstitch leaving only the back collar shape.

2. Select and delete the placket, buttons and back style lines leaving only the back bodice.

bring to front

1. Select the back collar with the **Selection** tool.

2. **Right-click** and choose **Arrange > Bring to Front.**

reshape

1. Use the **Direct Selection tool** to move the **CB neckline anchor** (bodice) upward so that it is hidden beneath the back collar shape.

create back sleeve placket

Sleeve plackets vary in shape and style. Ideally the sleeve placket (sleeve opening) is shown on the back view of the sleeve. Topstitch details and closures can be added using the Pen tool while the button can be copied from the front tailored shirt (Copy/ Paste).

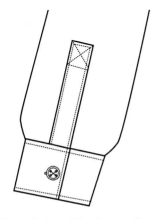

The back view of the sleeve cuff and placket with topstitch and button detail added.

QUICK TIP!
The Eyedropper tool allows you to sample stroke attributes. Practice using the Eyedropper in place of setting the stroke, fill and dashed line options.

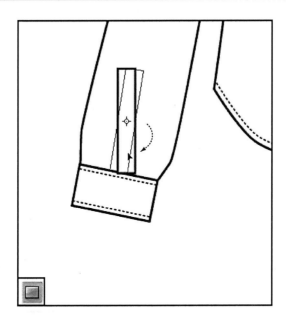

create rectangle

1. Select the **Rectangle** tool.

2. Drag diagonally to create a rectangle.

rotate

1. Select the **Rotate** tool.

2. Drag the rectangle to rotate it.

3. Use the Selection tool to place the rectangle at the cuff seam.

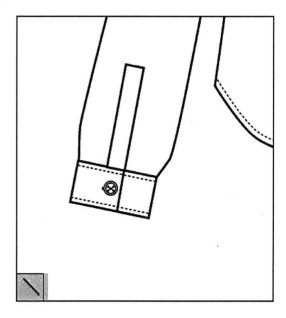

create line

1. Select the **Line Segment** tool.

2. Drag to create a line for the opening of the sleeve cuff.

duplicate button

1. Select the **Selection** tool.

2. Click to select a button on the front view of the tailored shirt.

3. Choose **Edit > Copy** then **Edit > Paste**.

4. Move the button over to the shirt cuff.

group details/ group items

Now that you have completed your tailored shirt's front and back views, you can group its seperate details. This will allow you to freely move fashion flats as one object, overlap them as in a presentation and easily ungroup them to explore new designs.

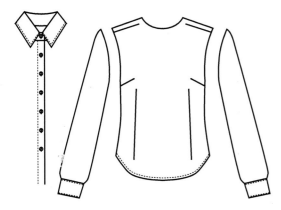

An ungrouped and exploded view of the tailored shirt with seperately grouped details.

> **QUICK TIP!**
> Be certain to save your work as often as every 5 to 10 minutes. A quick way to save is by using keyboard shortcuts. Choose File > Save or Ctrl+S. *(Mac) Cmd+S*

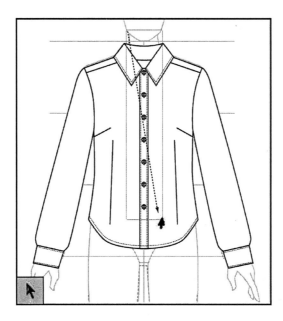

group front collar

1. Select the **Selection** tool.

2. Drag a selection marquee from outside the shirt then down and over to select the front collar parts, the placket and buttons.

   Note: The shirt bodice is included in the selection. You will use the Shift key to release it from the selection.

3. **Shift+click** the shirt bodice to release it from the selection.

   Note: Shift+click any other parts of the shirt that will not be included in the collar/ placket group.

4. Choose **Object > Group** to group the collar parts.

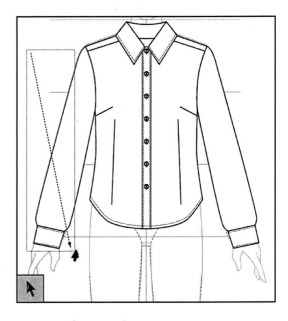

group front sleeves

1. With the the **Selection** tool drag from outside the shirt then over and down to select the left side sleeve parts.

2. Choose **Object > Group**.

3. With the the **Selection** tool drag from outside the shirt then over and down to select the right side sleeve parts.

4. Choose **Object > Group**.

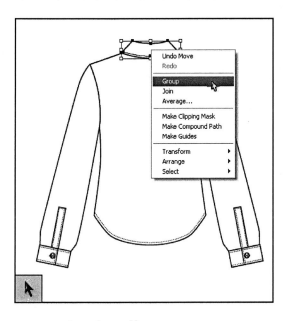

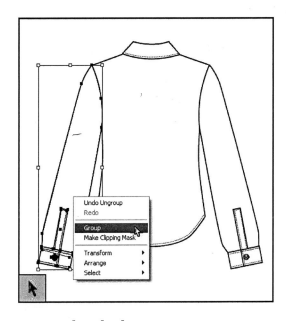

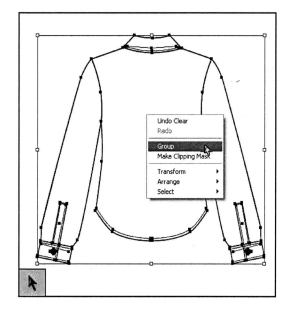

group back collar

1. With the the **Selection** tool drag from outside the shirt then over and down to select the back collar parts.

2. **Shift+click** the shirt bodice to release it from the selection.

3. Choose **Object > Group.**

group back sleeves

1. With the the **Selection** tool drag from outside the shirt then over and down to select the left side sleeve parts.

2. Choose **Object > Group.**

3. With the the **Selection** tool drag from outside the shirt then over and down to select the right side sleeve parts.

4. Choose **Object > Group.**

group entire front

1. With the the **Selection** tool drag from outside the shirt then over and down to select the tailored shirt front view.

2. Choose **Object > Group.**

group entire back

1. With the the **Selection** tool drag from outside the shirt then over and down to select the tailored shirt back view.

2. Choose **Object > Group.**

QUICK TIP!
A convenient way to choose the Group or Ungroup command is by using the onscreen context-sensitive menu. To do this select the desired objects then right-click on the artboard to choose Group. *(Mac) Control-click*

add color

Keep in mind that you only want to add color to closed shapes (bodice, sleeve, collar). If you add color to your open paths (topstitch, style lines) you will get unusual results that make parts of your sketch appear covered up. The buttons on the shirt are grouped and are all closed shapes. Use the Group Selection to select all the buttons then click the desired color in the Swatch panel.

Color added to topstitch appears incorrect.

Color added to the bodice, sleeve collar and buttons.

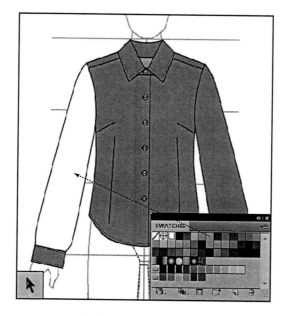

drag and drop color

1. Deselect all objects.

2. Click the **Swatches** panel.

3. Select the desired color then **drag and drop** the color into the shirt bodice.

drag and drop color

1. **Drag and drop** color into the collar parts (right, left, back and inside lining).

2. **Drag and drop** color to the sleeve parts (sleeve, cuff).

3. **Drag and drop** color to the back view parts of the tailored shirt.

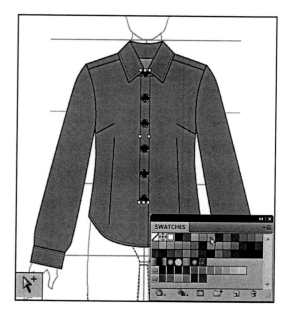

group select and color

1. Select the **Group Selection** tool.

2. Click a button on the front of the tailored shirt then click a second and third time so that all the buttons are selected.

3. Click the **Swatches** panel.

4. Click the desired color.

explore tailored shirt designs

Great! So now that you have completed the tailored shirt, you can explore tailored shirt designs. Begin with a copy of the shirt, ungroup it and remove details. Try your new tailored shirt with short sleeves or with an interesting collar.

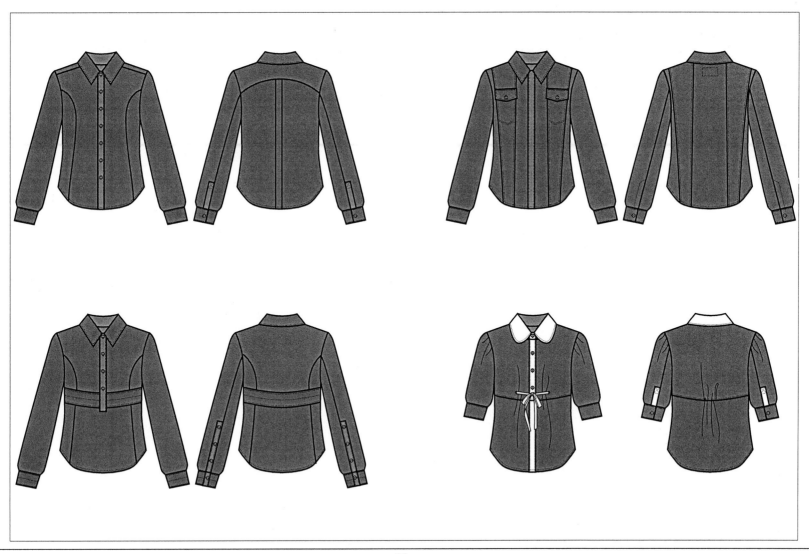

chapter seven

draw a denim pant
intermediate level lesson

So it is time to pick up the pace in this chapter. For this intermediate lesson it is assumed that you are familiar with how to select anchor points, convert anchor points, select and reflect, add to shape area and copy/paste. Also in this chapter keyboard shortcuts are used throughout to promote ease of use.

In this chapter you will create a boot cut, denim pant with western pockets and grommets, a zipper fly-front, buttoned waistband, belt loops and topstitch detail. As demonstrated in previous lessons a copy of the front view will aid in creating the back view.

To complete this denim pant you will add scanned denim fabric and then prepare the page for printing.

create pant shape

Pant shapes vary from cropped to long lengths and from wide to pencil shapes. Style lines and details vary on the style of pant and topstitch on denim is important to its construction. The task is to create the shape of the fashion flat and from there you can add style lines, topstitch and details such as a pocket or waistband. To start you will create the half pant shape setting the first anchor at the croqui's CF waist and the last anchor at the CF crotch. The shape is then reflected where it is important to remember to overlap the two shapes at the croqui CF.

Plan the half pant shape to determine the position of its anchors.

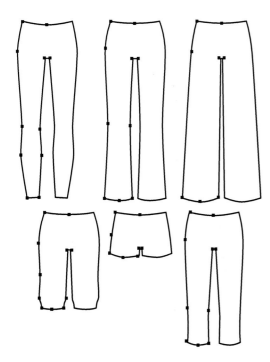

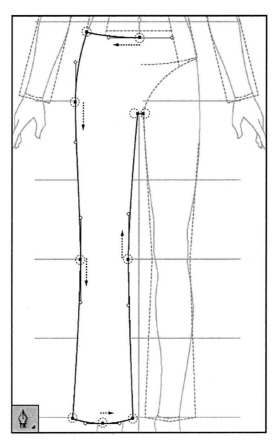

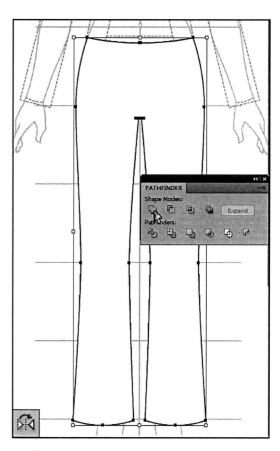

create pant shape

1. Click the **Default Fill and Stroke** in the **Tools** panel.

2. Draw the half pant with the **Pen** tool.

3. Smooth the pant shape with the **Convert Anchor Point** tool.

reshape

1. Reshape the pant shape with the **Direct Selection** tool.

reflect

1. Select the pant shape with the **Selection** tool.

2. Click the **Reflect** tool then hold the **Alt** key.

3. **Alt+click** at the CF with the **Reflect** tool.

unite

1. Select both shapes with the **Selection** tool.

2. Click **Unite** in the **Pathfinder** panel.

add topstitch

Topstitch on a denim pant is important to the design and as well the construction of the pant. Commonly you will find edge stitch and topstitch on styles where the fabric is heavier and/or more coarse.

Note the following stroke attributes:

Closed Shape
White fill/ black stroke

Style Line
None fill/ black stroke

Drape Line
None fill/ black stroke
Weight (0.5 pt) with Dashed Line option off.

Edge Stitch/ Topstitch
None fill/ black stroke
Weight (0.5 pt) with the Dashed Line option on. Type (1.0 pt) in each dash/ gap.

Small Detail
White fill/ black stroke
Weight (0.5 pt) with Dashed Line option off.

QUICK TIP!
The Alt key on your keyboard is used to make a duplicate of an object. Hold the Alt key while moving the object with the Selection tool to make its duplicate (Alt+drag). *(Mac) Option key*

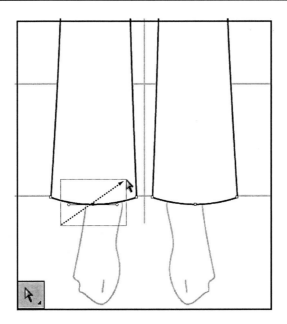

copy/ paste path

1. Select the left pant hem with the **Direct Selection** tool.

   Note: Do not include the side pant anchors in this selection.

2. Choose **Edit >Copy** then **Edit > Paste**.

3. Position and scale the path with the **Selection** tool.

4. Click **None** in the **Tools** panel.

5. Set the stroke **Weight** and **Dashed Line** in the **Stroke** panel.
   (Stroke Weight 0.5 pt/ Dashed Line 1 pt)

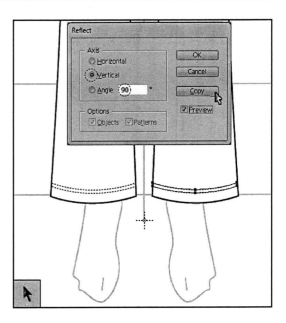

duplicate topstitch

1. Select the topstitch with the **Selection** tool then hold the **Alt** key.

2. **Alt+drag** the topstitch to duplicate it.

reflect

1. Select both topstitch paths with the **Selection** tool.

2. Click the **Reflect** tool then hold the **Alt** key.

3. **Alt+click** at the CF with the **Reflect** tool.

create western pocket

For the western pocket you will use the Arc tool to create the shape and topstitch. When creating small details like the button or rivet you will want to draw the detail oversized, group it and then scale it to fit the fashion flat. Grouping small details make the editing of fashion flats easier and more efficient. Once the small detail is scaled remember to set its stroke weight in the Stroke panel. Small details such as rivets look best with a 0.5 pt stroke weight.

Draw the rivet oversized then group and scale it to fit on the pant.

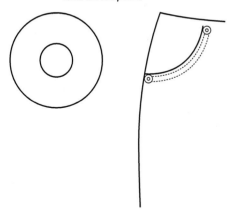

QUICK TIP!
The Shift key on your keyboard constrains the drawing, moving and transforming of objects. To create a proportioned shape hold the Shift key while creating the shape.

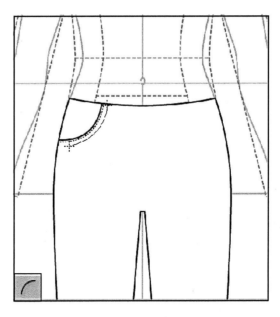

create western pocket

1. Click **Default Fill and Stroke** in the **Tools** panel.

2. Click the **Arc** tool then hold the **Shift** key.

3. **Shift+drag** diagonally to create the pocket.

add topstitch

1. Click the **Arc** tool then hold the **Shift** key.

2. **Shift+drag** to create the pocket topstitch.

3. Set the stroke **Weight** and **Dashed Line** in the **Stroke** panel.
   (Stroke Weight 0.5 pt/ Dashed Line 1 pt)

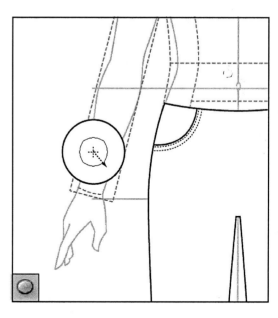

create rivet

1. Click **Default Fill and Stroke** in the **Tools** panel.

2. Click the **Ellipse** tool then hold the **Shift** and **Alt** keys.

3. **Shift+Alt+drag** to create a circle.

4. **Shift+Alt+drag** to create an inner circle.

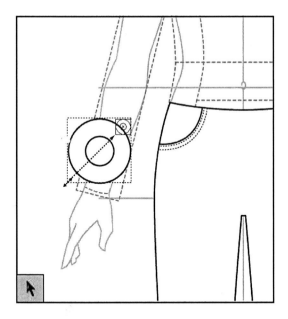

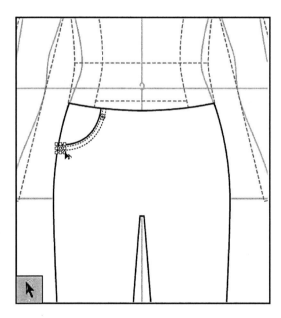

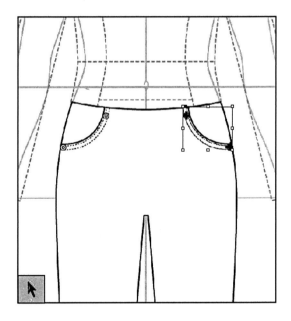

group and scale

1. Select both circles with the **Selection** tool.

2. Choose **Object > Group**.

3. **Shift+drag** the corner handle to scale the rivet.

4. Set the stroke **Weight** in the **Stroke** panel.
    (Stroke Weight 0.5 pt)

QUICK TIP!
Did you turn off the Scale Stroke and Effects in the Scale tool? Double click the Scale tool in the Tools panel and uncheck the option.

---

duplicate rivet

1. Position the rivet with the **Selection** tool.

2. **Alt+drag** the rivet to duplicate it.

group and reflect

1. Select the pocket, its topstitch and the rivets with the **Selection** tool.

2. Choose **Object > Group**.

3. Click the **Reflect** tool.

4. **Alt+click** at the CF with the **Reflect** tool.

5. Deselect with the **Selection** tool.

create fly-front

It is always beneficial to refer to an actual pair of denim pants when drawing pants. As you research the various styles of denim pants and pants in general you will discover all sorts of unique front closures. Most commonly denim pants will have either an encased fly-front or a buttoned fly-front with variations of topstitch detail.

Variations of the denim pant fly-front.

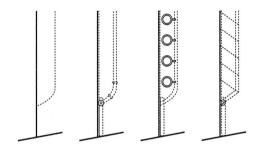

> **QUICK TIP!**
> The Eyedropper tool samples stroke and fill attributes. To sample with the Eyedropper tool select an object then eyedrop the object that you want to sample.

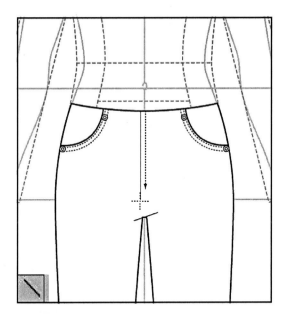

create fly-front

1. Click the **Line Segment** tool then hold the **Shift** key.

2. **Shift+drag** to create the pant closure.

3. Set the stroke **Weight** in the **Stroke** panel.
   (Stroke Weight 1 pt)

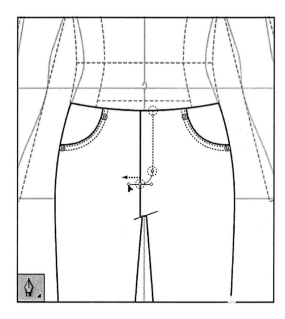

add topstitch

1. Draw the fly-front topstitching with the **Pen** tool.

2. Set the path's **Weight** and **Dashed Line** in the Stroke panel.
   (Stroke Weight 0.5 pt/ Dashed Line 1 pt)

duplicate topstitch

1. Click the **Selection** tool then hold the **Alt** key.

2. **Alt+drag** the topstitch to duplicate it.

3. Position and scale the topstitch with the **Selection** tool.

4. Deselect with the **Selection** tool.

create waistband

The waistband is created from a simple rectangle. Start with the rectangle then use the warp command to give the waistband a bit of shape (Object > Envelope Distort > Make with Warp). An important, in fact very important thing to remember about using the warp command is this: Once you have completed the warp command choose Object > Expand to complete the warp. Doing this will return the warped shape back into vector lines where you can reshape and add color.

Various warp Styles
applied to a rectangle.

Style: Arc

Style: Bulge

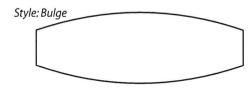

Style: Shell Upper

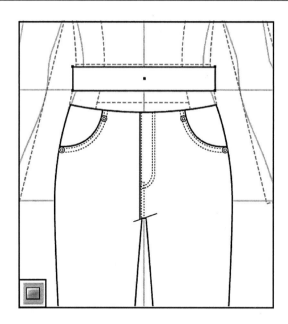

create waistband

1. Click **Default Fill and Stroke** in the **Tools** panel.

2. Draw a rectangle with the **Rectangle** tool.

QUICK TIP!
To easily move the artboard within the workspace (scroll) hold the Spacebar on your keyboard to get the Hand tool.

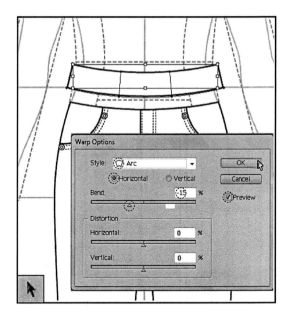

arc waistband (expand!)

Front

1. Select the rectangle with the **Selection** tool.

2. Choose **Object > Envelope Distort > Make with Warp**. Choose a warp **Style**. *(Arc)*

3. Click the **Horizontal** button then check the **Preview** button.

4. Drag the **Bend** slider backward to a minus number. *(Between -10% and -15%)*

5. Click **OK**.

6. Choose **Object > Expand**. Click **OK**.

7. Position then scale the waistband with the **Selection** tool.

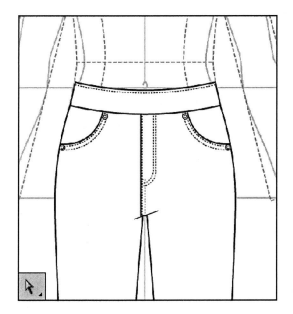

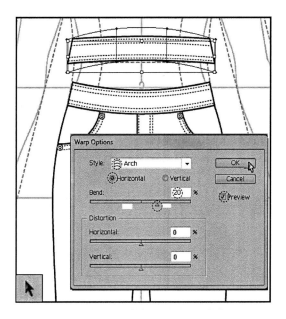

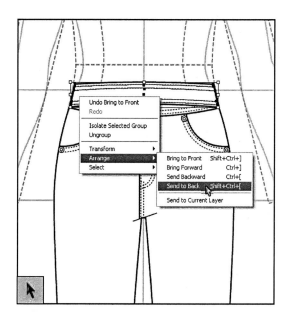

add topstitch

1. Select the top waistband path with the **Direct Selection** tool.

2. Choose **Edit >Copy** then **Edit > Paste**.

3. Position and scale the path with the **Selection** tool.

4. Click **None** in the **Tools** panel.

5. Set the stroke **Weight** and **Dashed Line** in the Stroke panel.
   (Stroke Weight 0.5 pt/ Dashed Line 1 pt)

duplicate topstitch

1. Click the **Selection** tool then hold the **Alt** key.

2. **Alt+drag** the topstitch to duplicate it.

arch waistband (expand!)

Back

1. Select the waistband and its topstitch with the **Selection** tool.

2. Choose **Edit >Copy** then **Edit > Paste**.

3. Choose **Object > Envelope Distort > Make with Warp**. Choose a warp **Style**. *(Arch)*

4. Click the **Horizontal** button then check the **Preview** button.

5. Drag the **Bend** slider forward to a plus number. *(Between **20%** and **25%**)*

6. Click **OK**.

7. Choose **Object > Expand**. Click **OK**.

send to back

1. Position then scale the waistband with the **Selection** tool.

2. Choose **Object > Arrange > Send to Back**.

create waistband closure

1. Click the **Line Segment** tool then hold the **Shift** key.

2. **Shift+drag** to create the pant closure.

3. Set the stroke **Weight** in the Stroke panel.
   (Stroke Weight 1 pt)

create button

Remember, creating a button is easy using the Ellipse and the Rectangle tool. You should draw the button and buttonhole oversize and away from the pant. By drawing the detail oversize you will have the ability to add subtle details such as the button groove and buttonhole topstitch. Group the completed button, scale it to the desired size then set its 0.5 pt stroke weight in the Stroke panel.

*The button is made from
an ellipse and a rectangle*

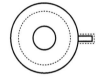

QUICK TIP!
You may find that you do not need the Croqui beyond this point. To hide the croqui, click the Layers panel then click the eye on the croqui layer.

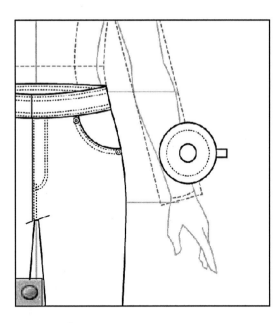

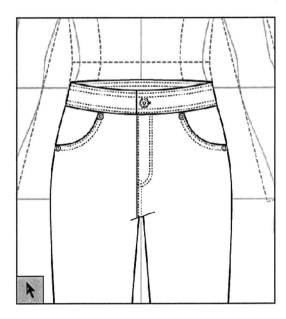

create button

1. Click **Default Fill and Stroke** in the **Tools** panel.

2. Click the **Ellipse** tool then hold the **Shift** and **Alt** keys.

3. **Shift+Alt+drag** to create a circle.

4. **Shift+Alt+drag** to create an inner circle.

create buttonhole

1. Draw a buttonhole with the **Rectangle** tool.

2. Choose **Object > Arrange > Send to Back**.

group and scale

1. Select the button and buttonhole with the **Selection** tool.

2. Choose **Object > Group**.

3. **Shift+drag** the corner handle to scale the button.

4. Set the stroke **Weight** in the Stroke panel. *(Stroke Weight 0.5 pt)*

5. Position the button with the **Selection** tool.

6. Deselect with the **Selection** tool.

create belt loop

Just like the rivet and the button on this pant you will create the belt loop with a simple shape tool, the Rectangle tool, drawing it oversize and away from the pant. Add the topstitch with the Line Segment tool, group the finished belt loop then scale it down to the desired size. Set the 0.5 pt stroke weight in the Stroke panel.

The belt loop is made from a rectangle and lines.

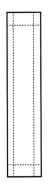

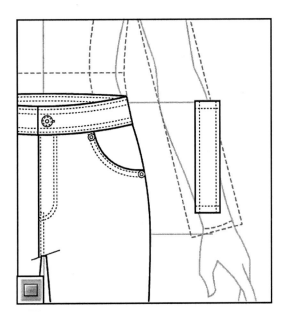

create belt loop

1. Click **Default Fill and Stroke** in the **Tools** panel.

2. Draw a belt loop with the **Rectangle** tool.

add topstitch

1. Click the **Line Segment** tool then hold the **Shift** key.

2. **Shift+drag** to create the topstitch.

3. Set the stroke **Weight** and **Dashed Line** in the Stroke panel.
   (Stroke Weight 0.5 pt/ Dashed Line 1 pt)

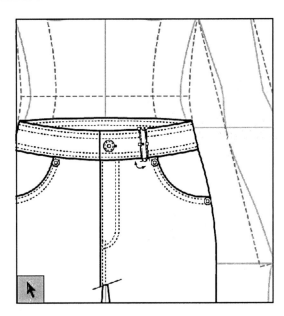

group and scale

1. Select the belt loop and its topstitch with the **Selection** tool.

2. Choose **Object > Group**.

3. **Shift+drag** the corner handle to scale the belt loop.

4. Set the stroke **Weight** in the Stroke panel.
   (Stroke Weight 0.5 pt)

rotate and reflect

1. Rotate the belt loop with the **Rotate** tool.

2. Position the belt loop with the **Selection** tool.

3. Reflect the belt loop with the **Reflect** tool.

create back view

Half the battle is won so now let's move on to the back view. Remember, the back view is a copy of the front view. The shape is the same, the details change. On the back view of the pant you'll have a CB seam and back yoke. To borrow details from the front view (belt loops/ button) be certain to use the Copy/ Paste command so that the details are stacked on top of the back view.

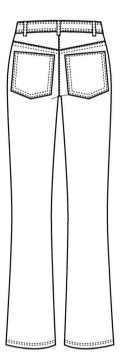

The pant back view is a copy of the front view with changes made to the CB, the pockets and the waistband.

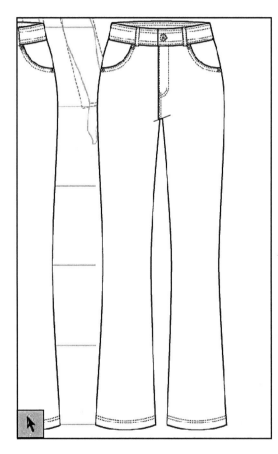

copy/ paste pant

1. Select the front view pant with the **Selection** tool.

2. Choose **Edit >Copy** then **Edit > Paste**.

3. Position the copied pant so that it is next to the original.

delete details

1. Select and delete details leaving only the back waistband and CB seam.

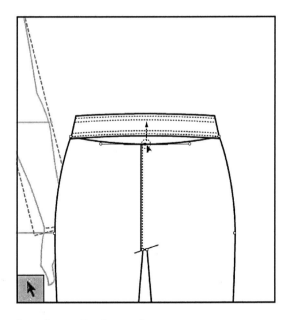

back waistband

1. Select the back waistband with the **Selection** tool.

2. Choose **Object > Arrange > Bring to Front.**

reshape

1. Reshape the back pant shape with the **Direct Selection** tool.

back belt loops

1. Select a belt loop with the **Selection** tool.

2. Choose **Edit >Copy** then **Edit > Paste.**

3. Position and rotate the belt loop on the back view.

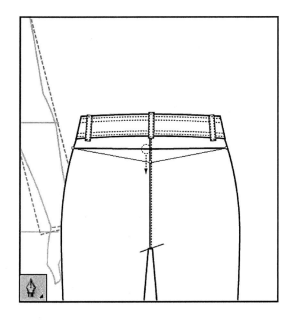

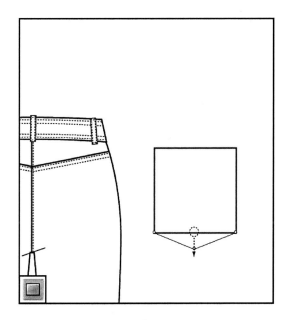

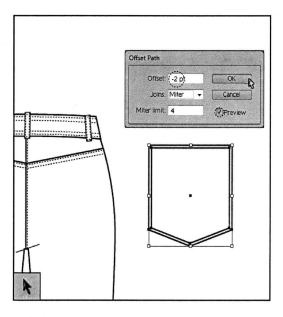

create back yoke

1. Draw the back yoke with the **Pen** tool.

   Note: Hold the **Shift** and click to set three anchors across the back pant. (left,center, right)

2. Reshape the yoke with the **Direct Selection** tool.

3. Click **None** in the **Tools** panel.

add topstitch

1. Click the **Selection** tool then hold the **Alt** key.

2. **Alt+drag** the yoke to duplicate it.

3. Set the stroke **Weight** and **Dashed Line** in the Stroke panel.
   (Stroke Weight 0.5 pt/ Dashed Line 1 pt)

create back pocket

1. Deselect with the **Selection** tool.

2. Click the **Default Fill and Stroke** option in the Tools panel.

3. Draw a rectangle with the **Rectangle** tool.

4. Click the **Add Anchor Point** tool.

5. Click on the bottom path to add an anchor.

6. Reshape the pocket with the **Direct Selection** tool.

add topstitch

1. Select the pocket shape.

2. Choose **Object > Path > Offset Path**.

3. Type **-2.0 pt** in the **Offset** field. Click **OK**.

4. Deselect then select the inner rectangle.

5. Click **None** in the **Tools** panel.

6. Set the stroke **Weight** and **Dashed Line** in the Stroke panel.
   (Stroke Weight 0.5 pt/ Dashed Line 1 pt)

   Note: To complete the pocket group it, scale it then position it on the back view pant. Use the Rotate tool to rotate the pocket then the Reflect tool (Alt+click) to reflect the pocket to the opposite side.

group details/ group items

Imagine having to sort through all the paths, lines, shapes, topstitch and details to try to take this pant apart. By grouping the details and then the entire fashion flat you make exploring design efficient and more fun too.

The possibilities are endless for this pant where you can easily remove details, create new details and have five to ten more pant designs in just minutes.

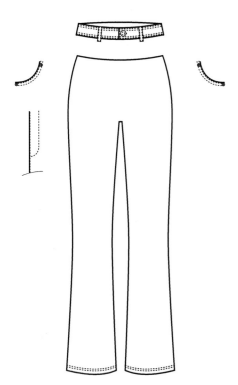

An exploded view of the pant shows separately grouped details.

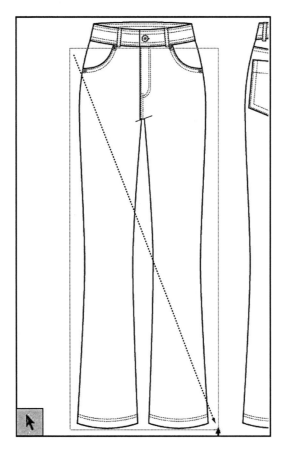

group front pant

1. Select the bottom part of the pant with the **Selection** tool.

   Note: Do not include the waistband in the selection.

2. Choose **Object > Group.**

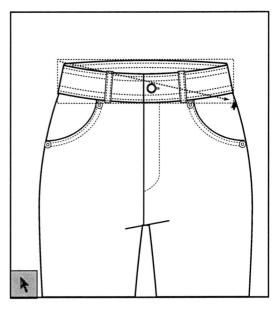

group front waistband

1. Select the front view waistband with the **Selection** tool.

   Note: The bottom part of the pant may be included in the Selection. Shift+click to release it from the selection.

2. Choose **Object > Group.**

group entire front

1. Select the entire front pant with the **Selection** tool.

2. Choose **Object > Group.**

group back view

So the rule is to group small details as you create them (buttons, belt loops, collars, pockets etc.) and then group the completed fashion flat front view and back view.

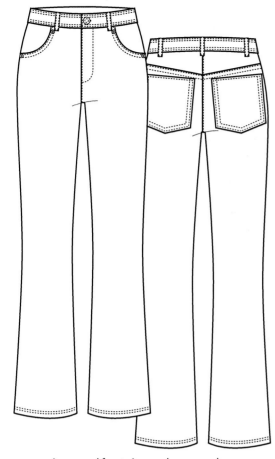

A grouped front view and a grouped back view makes it easier to overlap the fashion flat.

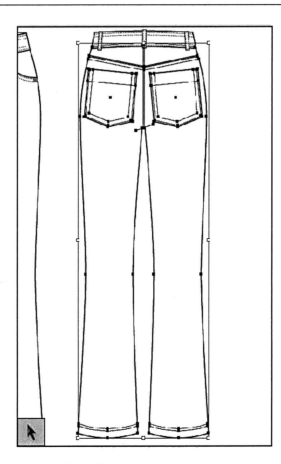

group back pant

1. Select the bottom part of the back view with the **Selection** tool.

2. Choose **Object > Group**.

3. Choose **Object > Arrange > Send to Back**.

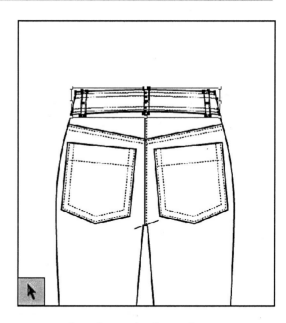

group back waistband

1. Select the back view waistband with the **Selection** tool.

   Note: The bottom part of the pant may be included in the Selection. Shift+click to release it from the selection.

2. Choose **Object > Group**.

group entire back

1. Select the entire back pant with the **Selection** tool.

2. Choose **Object > Group**.

add fabric

Included in the Woman Croqui document and in its Swatches panel is a collection of pre-scanned fabrics. Add denim fabric to your basic pant just as simply as you would add color.

The wrong way to add color or fabric is by selecting the entire fashion flat and then clicking the desired color. The best way to add color or fabric is by dragging the color over to the closed shape that makes up the outline of the fashion flat. For small details be certain to zoom in and use the Group Selection to select the detail and click color.

Color added to topstitch appears incorrect.

change stroke color

The most efficient way to change the stroke color is by choosing the stroke color options from the Control panel. Select a path then click the stroke color downward arrow to choose a color.

Click to choose a stroke color.

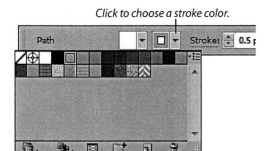

drag and drop fabric

1. Deselect with the **Selection** tool.

2. Click the denim fabric in the **Swatches** panel.

3. **Drag and drop** the denim fabric into the pant shape, waistband and belt loops.

group select and color

1. Click the **Group Selection** tool.

2. Click once then click a second time to select a rivet.

3. Click the desired color in the **Swatches** panel.

   Note: Repeat to color the remaining rivets and the button.

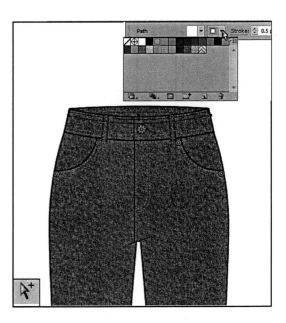

group select and stroke color

1. Click the **Group Selection** tool then hold the **Shift** key.

2. **Shift+click** to select the topstitch lines.

3. Click the stroke color options in the **Control panel**.

4. Click the desired stroke color.

explore basic pant designs

You've done this sort of thing before. Explore
pant designs with this one fashion flat and you'll
find that you are able to create several ideas in
just a few minutes. Change the pant waistband
or its pockets, add new seams and/ or topstitch
detail.

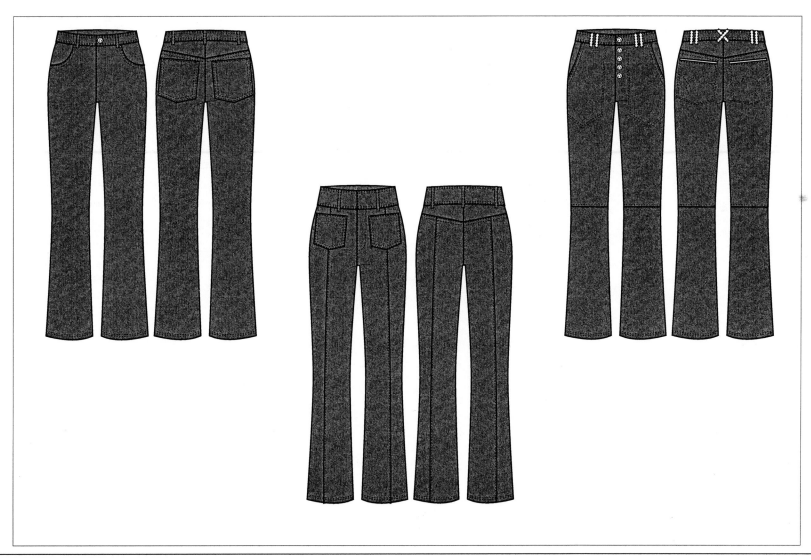

artboard tool

The Artboard tool in the Tools panel allows you to create, scale and position artboards within the workspace. Click the Artboard tool to view an artboard's bounding box, scale its size, position it within the workspace or create a new artboard.

The Artboard Options allow you to choose from a list of preset page sizes and set the page orientation. To view the Artboard Options double-click the Artboard tool then choose a Preset size (Letter or Tabloid) and Orientation (landscape or portrait).

> **QUICK TIP!**
> Be certain to save your work as often as every 5 to 10 minutes. A quick way to save is by using keyboard shortcuts. Choose File > Save or Ctrl+S.
> *(Mac) Cmd+S*

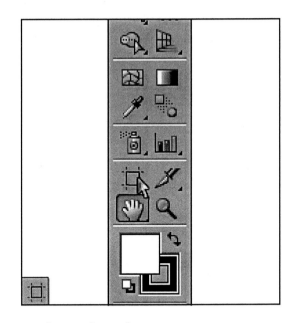

Double-click the Artboard tool to open the Artboard Options.

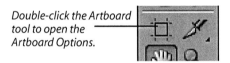

Choose an artboard Preset.

Choose an artboard Orientation.

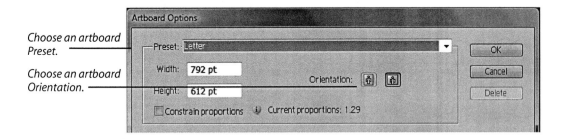

artboard options

1. Double-click the **Artboard** tool in the **Tools** panel.

2. Choose an artboard **Preset**. *(Letter)*

3. Choose an arboard **Orientation**. Click **OK**. *(landscape)*

4. Click the **Selection** tool to cancel the Artboard tool.

print setup

Once you have set the artboard's Preset size and Orientation in the Artboard Options you will then set the printer to print the page accordingly. Be certain the Show Print Tiling is on in the View menu then choose File > Print to open the Print dialog box. In the Print dialog box choose a Printer, a Media Size and Media Orientation.

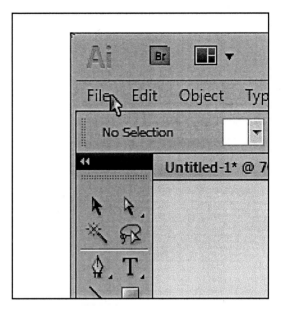

Choose File > Print to open the Print dialog box.

Choose a Printer.

Choose a Media Size.

Choose a Media Orientation.

Print preview

Print

Print Preset: Custom

Printer: Microsoft XPS Document Writer

PPD:

General

General
Marks and Bleed
Output
Graphics
Color Management
Advanced
Summary

Copies: 1 ☐ Collate ☐ Reverse Order
◉ All ☐ Ignore Artboards
○ Range: ☐ Skip Blank Artboards

Media
Size: Letter ☐ Transverse
Width: 612 pt Height: 792 pt

Options
Placement: X: 0 pt Y: 0 pt
◉ Do Not Scale
○ Fit to Page
○ Custom Scale: W: 100 H: 100
○ Tile Full Pages Overlap: 0 pt
☐ Scale: W: 100 H: 100
☐ Tile Range:

Print Layers: Visible & Printable Layers

1 of 1 (1)

Setup... Print Cancel Done

print setup

1. Choose **File > Print**.

2. Choose a **Printer.**

3. Choose a Media **Size**. *(Letter)*

4. Choose a Media **Orientation**. *(landscape 11" x 8.5")*

5. Click **Done**.

   OR

5. Click **Print**.

chapter eight

draw a flounce skirt
intermediate level lesson

The flounce skirt is the trickiest of all the fashion flats simply because you will want to create a hemline that simulates fabric drape. In this chapter you will create a flounce skirt with double topstitch at the hem and a wide waistband with a hidden zipper at the side seam. You will complete the front and back view of the skirt by adding a scanned floral fabric and then scale the size of the pattern to appear as a smaller or larger print. Finally you will use the Artboard tool to create a new artboard and then setup the printer to print each page.

make longer:

- Direct selection on bottom

- use arrows to move up & down

create skirt shape

Flounce skirts vary in detail, length and shape. For this flounce skirt you will create the general shape from the CF waist to the CF hemline. Create a zig-zag shape at the hemline that will be reshaped to create flounce. The Convert Anchor Point tool is then used to smooth the skirt shape. The Direct Selection tool is used to move anchors and direction handles and perfect the skirt shape.

Set the anchor points starting at the CF then around to create the zig-zag hem to the CF.

Use the Convert Anchor Point tool to smooth strategic anchor points. Smooth the hemline anchors toward the CF.

Perfect the shape with the Direct Selection tool moving anchors and direction handles.

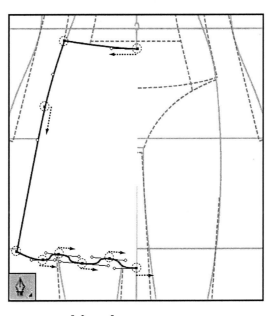

create skirt shape

1. Click the **Default Fill and Stroke** option in the **Tools** panel.

2. Draw the skirt shape with the **Pen** tool.

3. Smooth the skirt shape with the **Convert Anchor Point** tool.

reshape

1. Reshape the flounce hem with the **Direct Selection** tool.

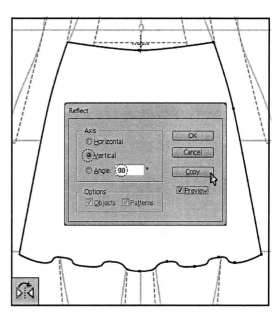

reflect

1. Select the skirt shape with the **Selection** tool.

2. Click the **Reflect** tool then hold the **Alt** key.

3. **Alt+click** at the CF with the **Reflect** tool.

unite

1. Select both shapes with the **Selection** tool.

2. Click **Unite** in the **Pathfinder** panel.

add topstitch

The topstitch on a flounce skirt is often times invisible unless it is a coarse fabric such as denim. The dash and gap of a dashed line can vary in size for example you may increase the gap and decrease the dash on a path to simulate blind stitch. Remember to always turn off the fill on topstitch lines (click None).

Recall the following stroke attributes:

Closed Shape
White fill/ black stroke

Style Line
None fill/ black stroke

Drape Line
None fill/ black stroke
Weight (0.5 pt) with Dashed Line option off.

Edge Stitch/ Topstitch
None fill/ black stroke
Weight (0.5 pt) with the Dashed Line option on. Type (1.0 pt) in each dash/ gap.

Small Detail
White fill/ black stroke
Weight (0.5 pt) with Dashed Line option off.

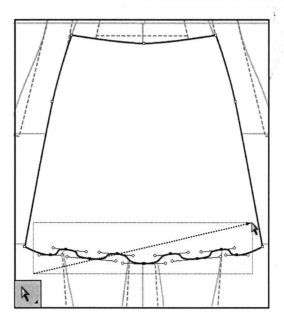

copy/ paste path

1. Select the skirt hem with the **Direct Selection** tool.

   Note: Do not include the side skirt anchors in this selection.

2. Choose **Edit >Copy** then **Edit > Paste**.

3. Position and scale the path with the **Selection** tool.

4. Click **None** in the **Tools** panel.

5. Set the stroke **Weight** and **Dashed Line** in the **Stroke** panel.
   (Stroke Weight 0.5 pt/ Dashed Line 1 pt)

duplicate topstitch

1. Select the topstitch with the **Selection** tool then hold the **Alt** key.

2. **Alt+drag** the topstitch to duplicate it.

3. Deselect with the **Selection** tool.

QUICK TIP!
To customize the look of your selections and anchor display choose Edit > Preferences > Selection and Anchor Display.

add drape lines

The easiest way to create drape lines on the skirt is to first create a line with the Line Segment tool then use the Convert Anchor point tool to smooth the drape line. Draw the line from the hemline then up into the skirt shape varying the length and direction of each. If you have mastered the Pen tool then you can easily draw in each drape line.

To use the Pen tool to create drape click at the hem to set the first anchor then drag at the second anchor to smooth the path.

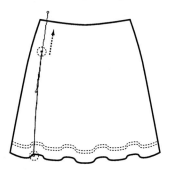

QUICK TIP!
When using the Pen tool create individual paths. Draw the first path by setting to anchors then press **P** on the keyboard to start a new path.

create drape lines

1. Set the drape line **Weight** in the Stroke panel.
   (Stroke Weight 0.5 pt/ Dashed Line Off)

2. Draw the skirt drape with the **Line Segment** tool.

3. Smooth the drape lines with the **Convert Anchor Point** tool.

add drape lines

1. Draw the drape line with the **Pen tool**.

2. Click **P** on the keyboard to start a new path.

3. Draw the next drape line with the **Pen tool**.

create waistband

The waistband is created from a simple rectangle and then an Arc warp is applied to give the waistband shape. Remember to expand the warp once you have completed it. This is a very important step that should never be overlooked (Object > Expand). The height of the waistband is dependant on the height of the rectangle that you initially create. For a high waistband create a high rectangle and for a slim waistband create a slim rectangle.

*Various waistband heights
with the Arc warp applied.*

Style: Arc (wide waistband)

Style: Arc (regular waistband)

Style: Arc (narrow waistband)

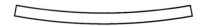

create waistband

1. Click the **Default Fill and Stroke** option in the **Tools** panel.

2. Draw a rectangle with the **Rectangle** tool.

arc waistband (expand!)

1. Select the rectangle with the **Selection** tool.

2. Choose **Object > Envelope Distort > Make with Warp**. Choose a warp **Style**. *(Arc)*

3. Click the **Horizontal** button then check the **Preview** button.

4. Drag the **Bend** slider backward to a minus number. *(Between **-10%** and **-15%**)*

5. Click **OK**.

6. Choose **Object > Expand**. Click **OK**.

7. Position then scale the waistband with the **Selection** tool.

create back waistband

The inside back of the waistband is a copy of the front waistband with an Arch warp applied to it. This shape is then sent to back and later a copy of it is used for the back view of the skirt.

A copy of the front waistband is Arch warped and then sent to back.

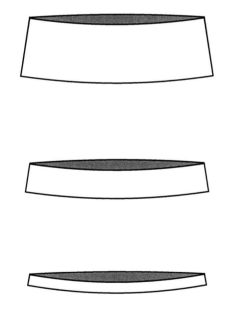

arch waistband (expand!)

1. Select the front waistband with the **Selection** tool.

2. Choose **Edit >Copy** then **Edit > Paste**.

3. Choose **Object > Envelope Distort > Make with Warp**. Choose a warp **Style**. *(Arch)*

4. Click the **Horizontal** button then check the **Preview** button.

5. Drag the **Bend** slider forward to a plus number. *(Between **20%** and **25%**)*

6. Click **OK**.

7. Choose **Object > Expand**. Click **OK**.

send to back

1. Position then scale the waistband with the **Selection** tool.

2. Choose **Object > Arrange > Send to Back**.

3. Deselect with the **Selection** tool.

create side closure

1. Draw the side topstitch with the **Pen** tool.

2. Turn off the path's fill in the **Tools** panel. *(Click None)*

3. Set the stroke **Weight** and **Dashed Line** in the Stroke panel.
   (Stroke Weight 0.5 pt/ Dashed Line 1 pt)

create back view

Can you believe how easy this skirt was? So now you can move on to the back view of the skirt. Not much has changed from other back views that you have created. The back view is a copy of the front view with small adjustments to the back waist.

The skirt back view is a copy of the front view with changes made to the back waist, side closure and back drape lines.

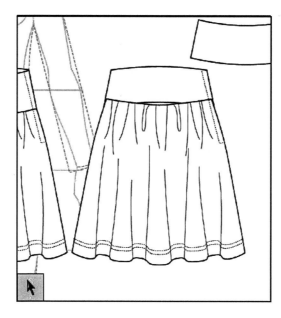

copy/ paste skirt

1. Select the front view skirt with the **Selection** tool.

2. Choose **Edit >Copy** then **Edit > Paste**.

3. Position the copied skirt so that it is next to the original.

delete details

1. Select and delete details leaving only the back waistband and side closure.

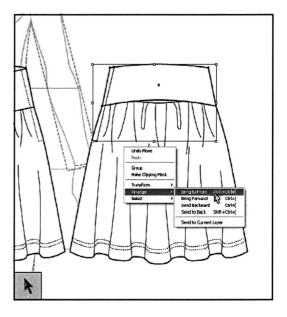

back waistband

1. Reflect the side closure to the opposite side of the skirt with the **Reflect** tool..

2. Select the back waistband and side closure with the **Selection** tool.

3. Choose **Object > Arrange > Bring to Front.**

reshape

1. Reshape the back skirt shape with the **Direct Selection** tool.

group details/ group items

It may seem tedious to group the seperate parts of your fashion flat but it really will make things easier when you are ready to explore flounce skirt designs. In fact you can now borrow details from other fashion flats you have created. With the skirt's waistband grouped seperately you can easily remove it in one step and then replace it with an alternate waistband. You can even borrow the waistband, pockets or fly-front you created from the denim pants in chapter seven.

The flounce skirt with details removed.

A new skirt with borrowed details.

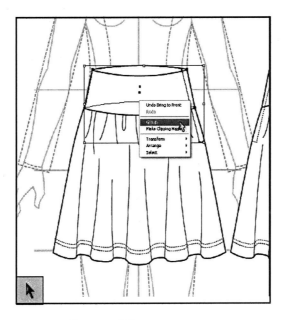

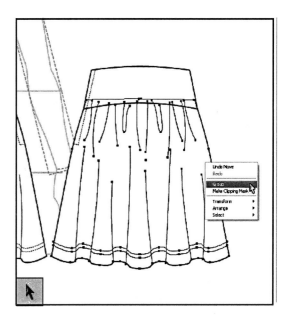

group front skirt

1. Select the front view skirt with the **Selection** tool.

2. Choose **Object > Group**.

group front waistband

1. Select the front view waistband with the **Selection** tool.

2. Choose **Object > Group**.

group entire front

1. Select the entire front skirt with the **Selection** tool.

2. Choose **Object > Group**.

group back skirt

1. Select the back view skirt with the **Selection** tool.

2. Choose **Object > Group**.

group back waistband

1. Select the back view waistband with the **Selection** tool. *(Drag a selection marquee)*

2. Choose **Object > Group**.

group entire back

1. Select the entire back skirt with the **Selection** tool.

2. Choose **Object > Group**.

add fabric

Here you will add a floral fabric to your flounce skirt.

Remember, drag the desired fabric over to the closed shape rather than selecting the entire skirt and clicking a color. This ensures that you are not adding color to your style lines, drape lines and topstitch.

scale fabric

You may choose to increase or decrease the size of the fabric print in your fashion flat. To do this first select the fashion flat with the Selection tool then double click the Scale tool. The Scale dialog box gives you the option to scale the pattern without changing the size of the object (sketch). Click the Preview button so that you are able to view the changes before clicking OK.

Click the Uniform button then type in a scale percentage.

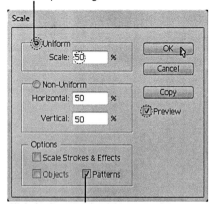

Click the Pattern check box to scale only the pattern.

drag and drop fabric

1. Deselect with the **Selection** tool.

2. Click the floral fabric in the **Swatches** panel.

3. **Drag and drop** the denim fabric into skirt shape and waistband.

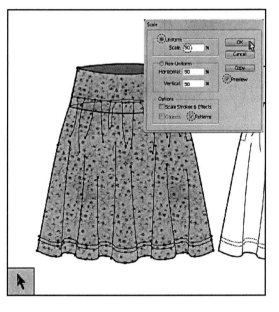

scale fabric

1. Select the skirt with the **Selection tool.**

2. Double-click the **Scale** tool in the Tools panel to open its dialog box.

3. Click the **Uniform** button then type in a number. *(Between 40% and 80%)*

4. Click the **Preview** button then uncheck the **Scale Stroke and Effect** button and the **Object** button.

5. Click **OK**.

explore flounce skirt designs

One way to re-design a flounce skirt is by
changing its length. Use the Direct Selection
tool to select the hem of the skirt then hold the
downward arrow to make the skirt longer or the
upward arrow to make it shorter.

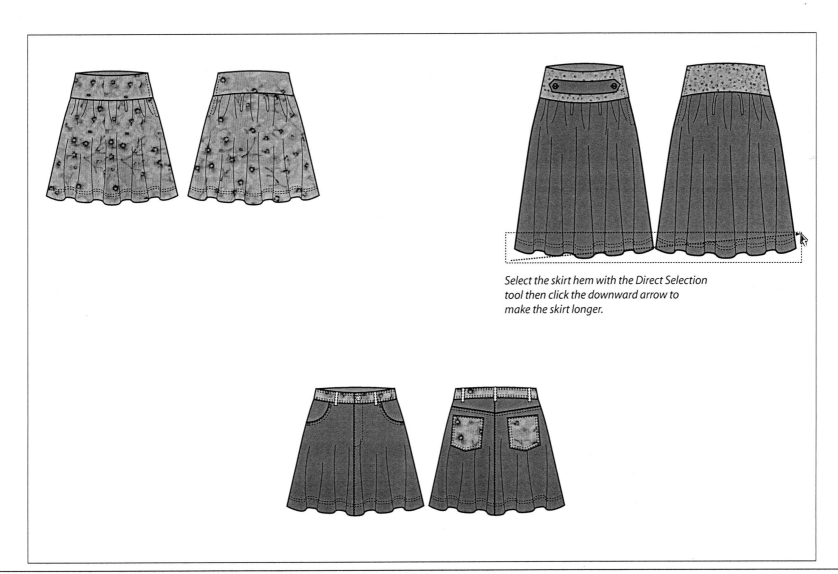

*Select the skirt hem with the Direct Selection
tool then click the downward arrow to
make the skirt longer.*

create new artboard

The Artboard tool allows you to create multiple artboards within one document where you can then scale, position and change the orientation of each artboard. Click the Artboard tool then drag diagonally within the workspace (away from the existing artboard) unitil the new artboard is the desired size (approximate the letter size). To set the size and orientation of the new artboard double-click to view the Artboard Options then choose a preset size and Orientation.

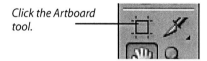

Click the Artboard tool.

Drag diagonally with the Artboard tool to create a new artboard.

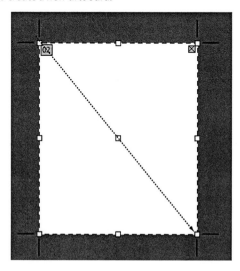

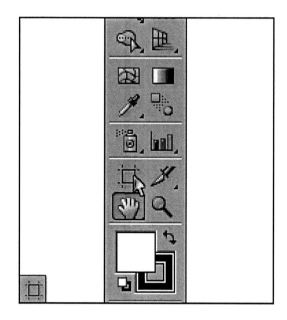

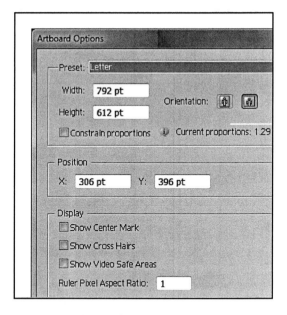

create new artboard

1. Click the **Artboard** tool in the **Tools** panel.

2. Draw a new artboad next to the existing one.

   Note: Approximate a letter size artboard.

artboard options

1. Double-click the **Artboard** tool in the **Tools** panel.

2. Choose an artboard **Preset**. *(Letter)*

3. Choose an arboard **Orientation**. Click **OK**. *(portrait)*

4. Click the **Selection** tool to cancel the Artboard tool.

   Note: Use the Artboard tool to position the new page next to the original.

print setup

Once you have set Artboard Options you will then set the printer to print each page accordingly. Choose File > Print to open the Print dialog box. In the Print dialog box choose a Printer, a Media Size and Media Orientation. You can also print all the pages or you can type in the range of pages that you would like to print. The print preview window allows you to click through the pages with its forward/ back arrows.

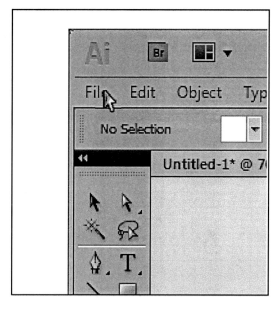

Choose File > Print to open the Print dialog box.

Choose a Printer.

Click to print All or Range.

Choose a Media Size.

Choose a Media Orientation.

Print preview

Click to view each page.

Print

Print Preset: Custom

Printer: Microsoft XPS Document Writer

PPD:

General

General
Marks and Bleed
Output
Graphics
Color Management
Advanced
Summary

Copies: 1   ☐ Collate   ☐ Reverse Order
◉ All   ☐ Ignore Artboards
◯ Range: _____   ☑ Skip Blank Artboards

Media
Size: Letter   ☐ Transverse
Width: 8.5 in   Height: 11 in

Options
Placement: X: 0 in   Y: 0 in
◉ Do Not Scale
◯ Fit to Page
◯ Custom Scale: W: 100   H: 100
◯ Tile   Full Pages   Overlap: 0 in
   ☐ Scale: W: 100   H: 100
   ☐ Tile Range:

2 of 2 (2)

Print Layers: Visible & Printable Layers

Setup...   Print   Cancel   Done

print setup

1. Choose **File > Print**.

2. Choose a **Printer.**

3. Click **All** to print all pages or click **Range** then type in the page numbers to print.

4. Choose a Media **Size**. *(Letter)*

5. Choose a Media **Orientation**. *(landscape 11" x 8.5")*

6. Click **Done**.

   OR

6. Click **Print**.

chapter nine

draw details and trim
intermediate level lesson

Thus far you have created a series of fashion
flats and a few details that accompany them
and you are familiar with exploring designs and
creating variations of the particular fashion flat.
In this chapter you will discover unique and
efficient ways to create detail and trim. Details
on a garment are often times the driving force
behind a design or concept. A simple fashion
flat whether it is a t-shirt or pant can be more
interesting when the right details are added to
them. Following is a list of details to consider
when designing fashion flats.

Pockets
Belts
Buckles
Belt Loops
Buttons
Elastic Waistbands
Drawstrings
Tie String
Bows
Overlock stitiching
Ric Rac trim
Ruffles
Scalloped Edge
Fur texture
Lacing
Ruffle tim

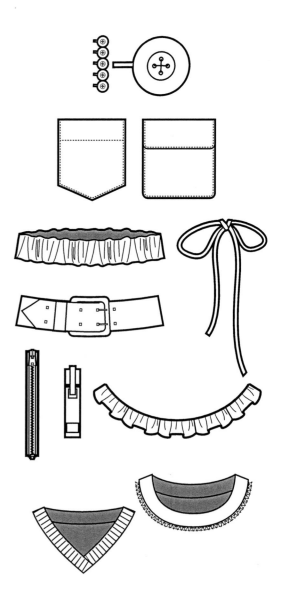

shape tools

Rectangle

The Rectangle tool is the ideal shape tool for creating pockets, tabs, buttonholes, waistbands, belts and buckles, belt loops, zipper pulls and zipper casings. Create a simple shape then add or delete anchors, apply a warp and/or reshape to create complex details and trim.

Rounded Rectangle

The Rounded Rectangle alternately can be used to create rounded pockets, buckles, buttons and various rounded rectangular shapes. Try using the various reshape tools such as the Knife tool or Eraser tool to easily create interesting details for your fashion flats.

Ellipse

The Ellipse tool is essential to creating buttons, grommets, rivets, and various circular details. Most details are created from an array of shape tools, sometimes requiring several circles or a circle and a rectangle combined. Such is the case with a button which is made up of a few circles, a rectangle and lines.

The Rectangle tool and its related shape tools.

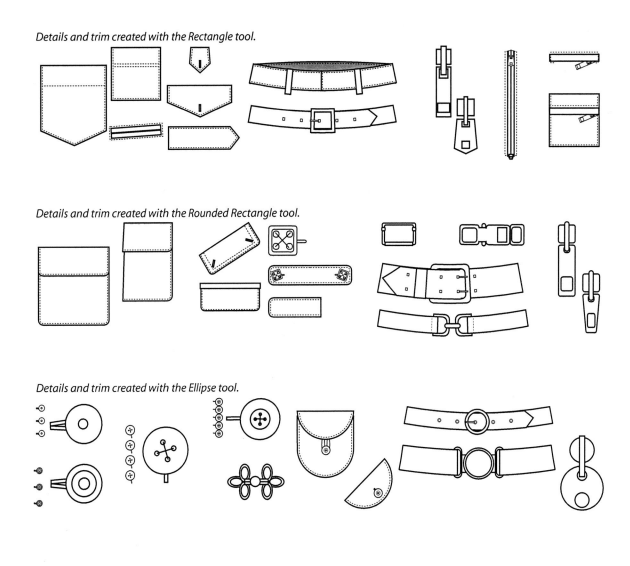

Details and trim created with the Rectangle tool.

Details and trim created with the Rounded Rectangle tool.

Details and trim created with the Ellipse tool.

create belt buckle

Simple belt buckles can be created using any of the shape tools in the Tools panel. The belt strap resembles the basic waistband which is created from a warped rectangle. The eyelets, prong and belt loop is created with the Ellipse tool and Rectangle tool. For more complex belt buckles, several shapes can be combined or divided. Use the Add to shape area command in the Pathfinder panel to combine several shapes. Use the Subtract from shape area command to divide shapes.

The belt buckle is created from two overlapping shapes and the Subtract from shape area command.

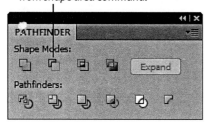

Select the two shapes then click Subtract from shape area, Expand.

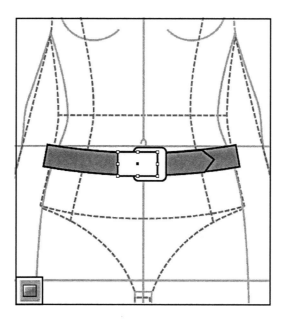

create belt strap

1. Create the basic waistband. *(see page 89)*

2. Draw the belt tip with the **Pen** tool.

create buckle shape

1. Click **Default Fill and Stroke** in the **Tools** panel.

2. Draw the buckle outline with the **Rounded Rectangle** tool.

2. Draw an overlapping shape with the **Rectangle** tool.

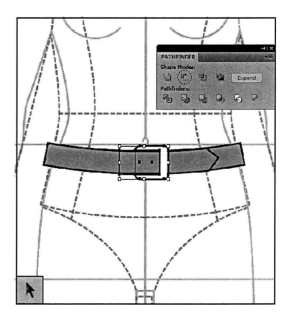

subtract from shape area

1. Select the belt buckle shapes with the **Selection** tool.

2. Click **Subtract from shape area** in the **Pathfinder** panel.

create zipper pull

For this very simple zipper pull you will use the Rectangle tool to create five overlapping shapes. This technique works with the Rectangle tool, the Rounded Rectangle and the Ellipse tool. To divide the shapes on the zipper handle use the Subtract from shape area command in the Pathfinder panel. Ideally you will want to draw this tiny detail oversize, scale it to fit the fashion flat and then set its stroke Weight to 0.5 pt in the Stroke panel.

The zipper pull is created with the Rectangle tool, the Rounded Rectangle tool and/or the Ellipse tool.

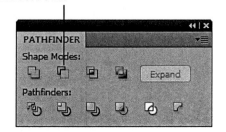

Use the Subtract from shape area command to make the zipper handle see-thru.

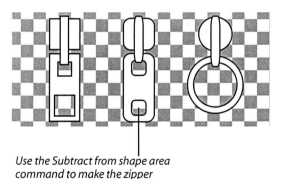

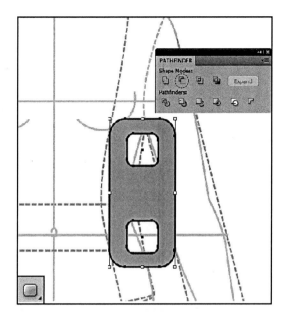

create zipper shapes

1. Click **Default Fill and Stroke** in the **Tools** panel.

2. Draw the zipper handle with the **Rounded Rectangle** tool.

3. Draw two more rectangles on top of the handle.

subtract from shape area

1. Select the three rectangles with the **Selection** tool.

2. Click **Subtract from shape area** in the **Pathfinder** panel.

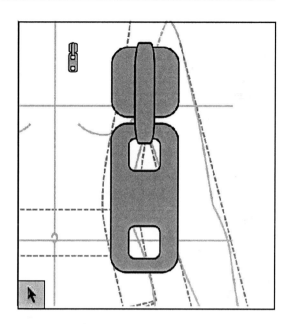

create zipper shapes

1. Draw the zipper head with the **Rounded Rectangle** tool.

2. Draw a rectangle to connect the zipper head to its handle.

group and scale

1. Select the complete zipper pull with the **Selection** tool.

2. Choose **Object > Group**.

3. **Shift+drag** the corner handle to scale the zipper pull.

4. Set the stroke **Weight** in the Stroke panel. *(Stroke Weight 0.5 pt)*

stroke panel

You have used the Stroke panel up to this point to set stroke weight and dashed line options (topstitcn) for your fashion flats. In additon to creating topstitch you can also use the stroke panel to create such details as zipper teeth and ribbing. Stroke weights can vary from 0.5 pt to 100 pt and dashed line options can vary in gap and dash size. To create zipper teeth first create the path, apply a thick stroke weight (from 1 pt to 3 pt) then apply a dashed line (dash: 1 pt and gap: 1 pt).

Details, trim and texture created with the Stroke panel.

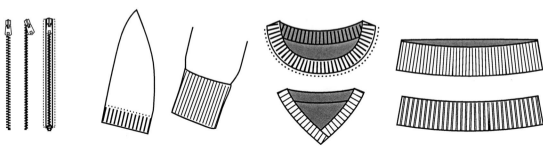

Create zipper teeth with two thick weight dashed lines.

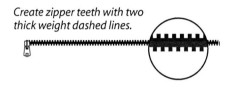

Stroke Weight/ Dashed Line variations:

weight: 1. pt (no dash)

weight: 3 pt / dash: 1 pt / gap: 1 pt

weight: 10 pt / dash: .5 pt / gap: 2 pt

weight: 12 pt (no dash)

weight: 14 pt / dash: 1 pt / gap: 2 pt

weight: 18 pt / dash: .5 pt / gap: 5 pt

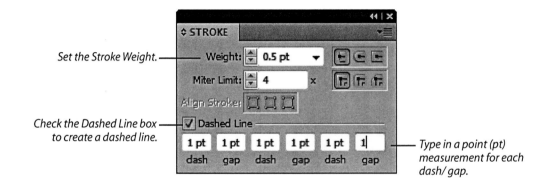

Set the Stroke Weight.

Check the Dashed Line box to create a dashed line.

Type in a point (pt) measurement for each dash/ gap.

create ribbing

To create ribbing for example on a neckline, sleeve cuff or waistband you should first create the shape that will contain the rib texture. Ideally you can copy/ paste a path from the outline shape and then position and scale to fit in the middle of the shape. From there you simply increase the stroke weight until the path fills in the shape then apply a dashed line to render the rib texture.

Ribbing is created from a path that fits in the center of the outline shape.

Use the Stroke panel to increase the stroke weight and set the dashed line option.

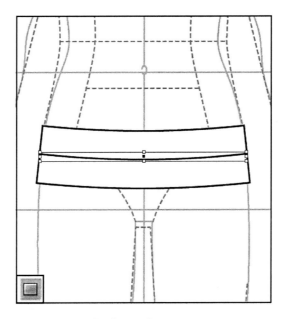

create waistband

1. Create the basic waistband.

copy/ paste path

1. Select the bottom path of the waistband with the **Direct Selection** tool.

2. Choose **Edit > Copy then Edit > Paste.**

3. Click the **Selection** tool.

3. Position and scale the path so that it fits in the center of the waistband.

4. Click **None** in the **Tools** panel.

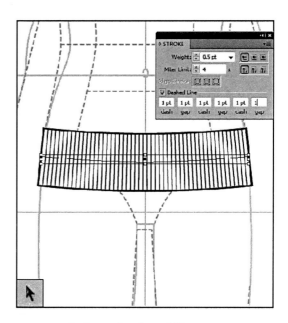

set weight/ dashed line

1. Set the stroke **Weight** in the Stroke panel. *(Stroke weight approximately 32 pt)*

   Note: Increase the stroke weight until the path fills up the waistband shape.

2. Set the **Dashed Line** in the Stroke panel. *(dash .5 pt/ gap 2 pt)*

liquify tools

Warp

The Warp tool is a liquify tool that warps a selected shape with a brush. Liquify tools are freehand tools that should not be confused with the Warp filter in the Object > Envelope Distort menu. The Warp tool is ideal to use when creating gathers in a shape as in fabric being pulled and gathered within the fashion flat. Such is the case with drawstrings, elastic and wrinkles. The Warp tool is used to create gather in the shape and then the Pen tool is used to draw in drape lines.

Experiment with other liquify tools to get a handle on how the tools work. To change the options of a liquify tool (brush dimensions, options) double-click the desired liquify tool in the Tools panel to open the Warp Tool Options.

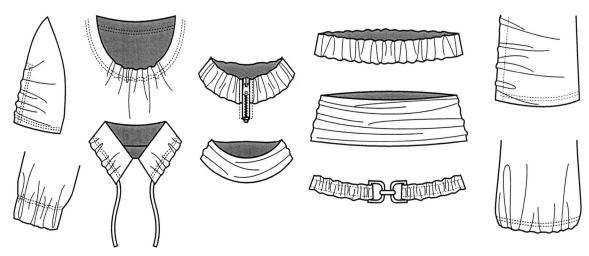

Details and trim enhanced with the Warp tool.

The Width tool, Warp tool and its related liquify tools.

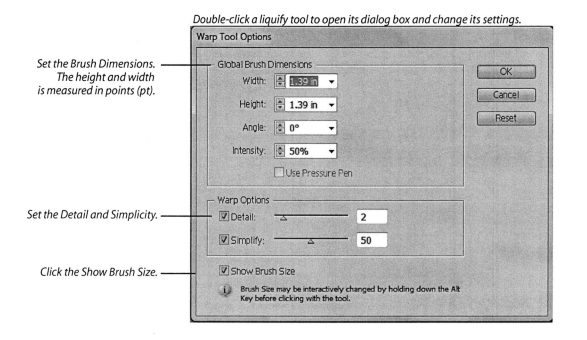

Double-click a liquify tool to open its dialog box and change its settings.

Set the Brush Dimensions. The height and width is measured in points (pt).

Set the Detail and Simplicity.

Click the Show Brush Size.

create elastic waistband

When creating a basic waistband remember to click the Default Fill and Stroke option in the Tools panel and then create a rectangle. To give the waistband a more natural shape apply the Make with Warp command and remember to Expand the warped shape *(See page 89)*.

The Warp tool is used to create a gathered effect. You will set the Warp tool options to a small brush size then use it to brush the gathers onto the waistband shape. To complete the elastic waistband the Pen tool is used to add drape lines. It is always beneficial to group the completed detail making it easier to edit and use on other fashion flats.

A flounce skirt with an elastic waistband.

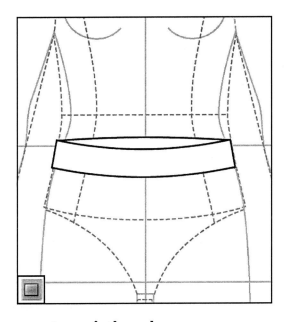

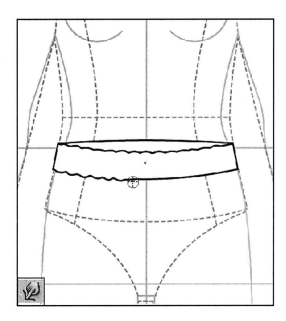

create waistband

1. Create the basic waistband. *(See page 89)*

warp tool options

1. Double-click the **Warp** tool in the Tools panel.

2. Set the **Brush Dimensions** in the Warp Tool Options dialog box. *(Width* **5 pt***, Height* **5pt***, Angle* **0***, Intensity* **60%***)*

3. Check the **Details** button then drag its slider. *(Between* **1** *and* **3***)*

4. Check the **Simplify** button then drag its slider. *(Between* **40** *and* **50***)*

5. Click **OK**.

create gather

1. Select the front waistband with the **Selection** tool.

2. Brush over the front waistband to create gather with the **Warp** tool.

3. Select the back waistband with the **Selection** tool.

4. Brush over the back waistband to create gather with the **Warp** tool.

5. Deselect with the **Selection** tool.

6. Draw the elastic drape lines with the **Pen** tool.

offset path

The Offset path command creates a duplicate outline of a selected object. If the object is a closed shape the offset path will create a duplicate shape either inside (minus before the number) or outside the original object. This is a perfect solution for adding topstitch to a pocket. If the object is an open shape the offset path will create an outline of the original object. Outlining an open shape is a solution for creating a tie or string for your fashion flat.

The Offset Path dialog box allows you to set an offset measurement. Depending on the size of the object you are offsetting the measurement will vary. For example you will create a pocket shape then offset the path by (-2 pt) for edge stitch or (-3 pt) for topstitch.

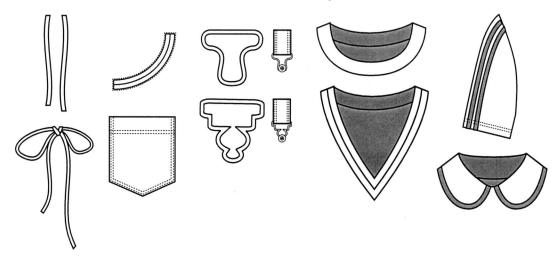

Details and trim created with the Offset Path command in the Object > Path menu.

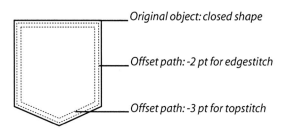

Original object: closed shape

Offset path: -2 pt for edgestitch

Offset path: -3 pt for topstitch

Choose Object > Path > Offset Path to open its dialog box.

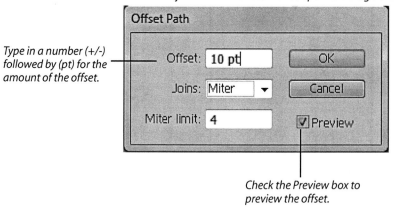

Type in a number (+/-) followed by (pt) for the amount of the offset.

Offset Path

Offset: 10 pt   OK

Joins: Miter   Cancel

Miter limit: 4   ☑ Preview

Check the Preview box to preview the offset.

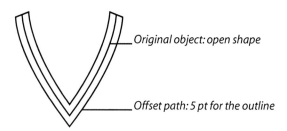

Original object: open shape

Offset path: 5 pt for the outline

create tie string

Here you will use the Offset Path command to create a tie string. The first task is to create the the shape of the tie. The finished tie will be a series of closed shapes that overlap to look like a tied string. Remember to click the Default Fill and Stroke option before drawing the shapes. You will create the open shapes with the Pen tool then use the offset path to outline the shapes.

Ideally you will want to draw the tie string detail oversize then scale it to fit the fashion flat. Remember to group the detail once you have completed it and also set the stroke weight to 0.5 pt in the Stroke panel.

A flounce skirt with a tie string waistband.

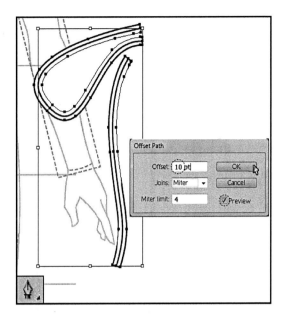

create tie string shape

1. Create the tie string shapes with the **Pen** tool.

offset path

1. Select the shapes with the **Selection** tool.

2. Choose **Object > Path > Offset Path** from the menu bar. Click the **Preview** button.

3. Type in a number for the **Offset** and include (pt) after the number for example (**3 pt**). Click **OK**.

4. Deselect then delete the original tie string shapes.

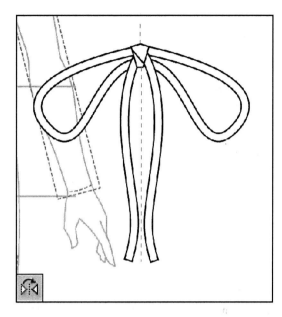

reflect

1. Select the tie string shapes with the **Selection** tool.

2. Click the **Reflect** tool then hold the **Alt** key.

3. **Alt +click** the center with the **Reflect** tool.

4. Create the knot detail with the **Pen** tool.

group and scale

1. Select the tie string with the **Selection** tool.

2. Choose **Object > Group**.

3. **Shift+drag** the corner handle to scale the tie.

4. Set the stroke **Weight** in the Stroke panel. *(Stroke Weight 0.5 pt)*

zig zag

The Zig Zag command in the Effects menu allows you to change a path into a zig zagged path. This command is usefull when creating overlock stitch or ric-rac trim. The Zig Zag dialog box lets you choose the size, the amount of ridges and either smooth or corner points for the zig zag. For complex zig zag detail you may opt to create the zig zag path and then create a Pattern Brush to be used with the Paintbrush tool. See page 123 for more information on creating a Pattern Brush.

Details and trim created with the Zig Zag command in the Effects > Distort & Transform menu.

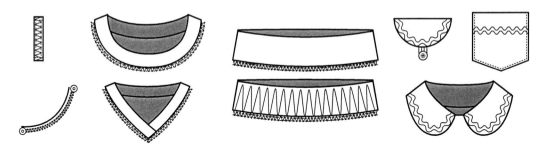

Zig Zag Line variations:

Size: 4 pt / Ridges per segment: 40 / Points: Smooth

Size: 8 pt / Ridges per segment: 40 / Points: Smooth

Size: 4 pt / Ridges per segment: 20 / Points: Corner

Size: 2 pt / Ridges per segment: 80 / Points: Corner

Size: 2 pt / Ridges per segment: 10 / Points: Corner

Choose Effects > Distort & Transform > Zig Zag to open its dialog box.

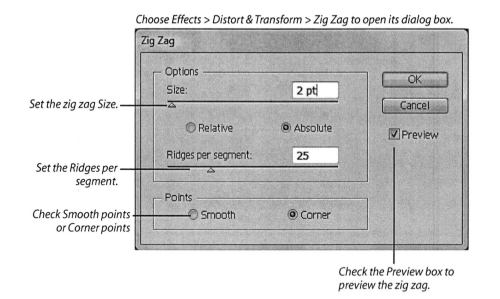

Set the zig zag Size.

Set the Ridges per segment.

Check Smooth points or Corner points

Check the Preview box to preview the zig zag.

create overlock stitch

Overlock stitch in clothing is defined as a reinforcement stitch inside of the garment that secures a seam or the raw edge of the fabric. Overlock can aslo be used decoratively on the outside of the garment. To create overlock for a fashion flat you will combine two topstitch lines and one zig zag line. To create overlock for a hemline copy/ paste the hemline path with the Direct Selection tool then apply topstitch attributes in the Stroke panel. Paste a second line for double topstitch then paste a third line to apply the zig zag command. Remember to turn off the fill on all topstitch detail including overlock stitch. To do this click the None option in the Tools panel.

Create overlock stitch with two topstitch lines and one zig zag line.

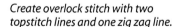
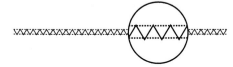

A basic waistband with overlock stitch.

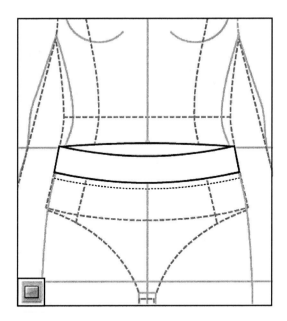

create waistband

1. Create the basic waistband.

create topstitch

1. Select the bottom path of the waistband with the **Direct Selection** tool.

2. Choose **Edit > Copy then Edit > Paste.**

3. Click **None** in the **Tools** panel.

4. Set the stroke **Weight** and **Dashed Line** in the Stroke panel.
   (Stroke Weight 0.5 pt/ Dashed Line 1 pt)

   Note: Duplicate the topstitch to create double topstitch.

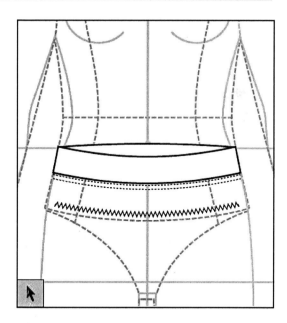

zig zag

1. Choose **Edit > Paste** to paste a copy of hte waistband path.

2. Set the stroke **Weight** and **Dashed Line** in the Stroke panel.
   (Stroke Weight 0.5 pt/ Dashed Line Off)

3. Choose **Effects > Distort & Transform > Zig Zag.**

4. In the **Zig Zag** dialog box check the **Preview** button.

5. Drag the **Size** slider to the desired position then drag the **Ridges per segment** slider to the desired position.

6. Click the **Corner** points button. Click **OK.**

7. Position the overlock and topstitch with the **Selection** tool.

brushes panel

The Brushes panel offers a selection of brush strokes that can be used to enhance your fashion flats. You can also create and edit unique brush styles to add details such as ruffles, scalloped trim, fur textures or frayed edges.

Of the four brush styles available in the Brushes panel (Calligraphic, Art, Scatter, Pattern) the Pattern Brush is most commonly used to add detail to fashion flats. To create a pattern brush you will first draw a horizontal tile (repeating detail) using various drawing tools then drag it into the brushes panel. When the Paintbrush is used the tile repeats along the path or paintbrush stroke.

Details, trim and texture (Pattern Brushes) created with the Brushes panel.

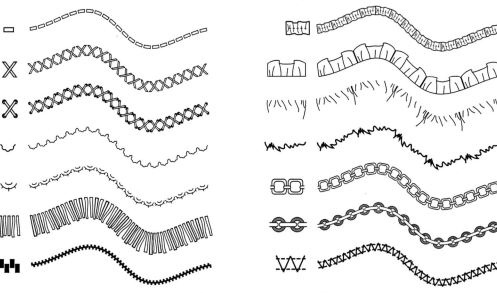

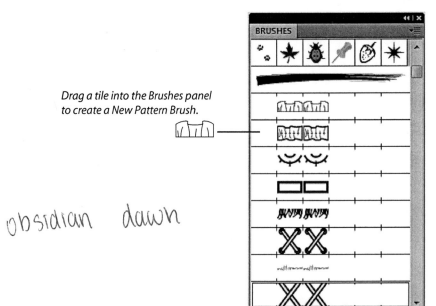

Drag a tile into the Brushes panel to create a New Pattern Brush.

Create one horizontal tile (repeating detail).

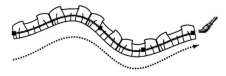

Create a pattern brush that will repeat the tile along a path or paintbrush stroke.

A scoop neckline with ruffle detail.

obsidian dawn

pattern brush

When creating the tile for your pattern brush there are three rules to keep in mind. The first is, when creating the tile draw it horizontally so that it will tile end by end along the path. Be sure that it is flat on either side so as to not leave a space when it tiles along the path. The second rule is, draw the tile as a closed shape with a white fill. This will ensure that the detail or trim will not be see through when you add it to your fashion flat. The third rule is to draw the detail oversize then scale it down to the actual size and stroke weight that you want it to appear as on your fashion flat.

The New Brush dialog box appears when you drag the tile into the Brushes panel.

Click to create a New Pattern Brush.

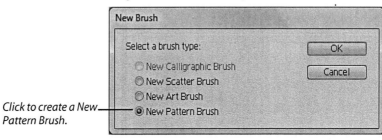

Draw the tile horizontally making sure it is flat on either side. Draw the tile as a closed shape with a white fill.

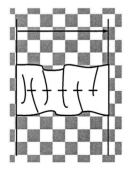

Type in a name for the new pattern brush.

The side tile will repeat along the path or the paintbrush stroke.

To control the brush color by setting the stroke color set the Colorization Method to "Hue Shift".

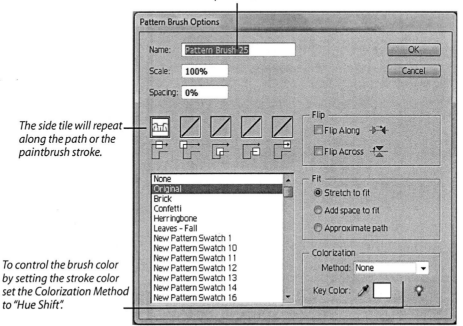

create ruffle pattern brush

To edit a pattern brush once it has been created double-click the brush in the Brushes panel to open the Pattern Brush Options dialog box then make the desired changes. Another way to make changes to your pattern brush is by changing the path's stroke Weight using the Stroke panel.

Setting the stroke to a thicker/ thinner weight will change the size of the brush.

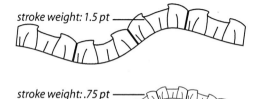

stroke weight: 1.5 pt

stroke weight: .75 pt

Because a pattern brush does not reflect when the path is reflected it is ideal to create two versions of the pattern brush.

Create two versions of the pattern brush for left/ right details.

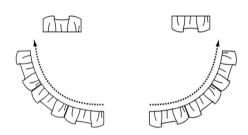

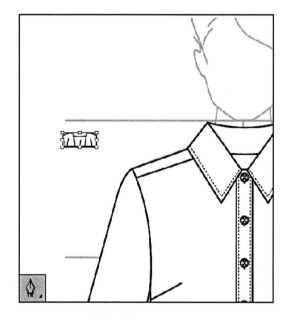

create ruffle tile

1. Create the ruffle tile and its drape lines with the **Pen** tool and **Line Segment** tool.

2. Select the entire tile with the **Selection** tool then hold the **Shift** key.

3. **Shift+drag** the corner handle to scale the tile.

4. Set the stroke **Weight** in the Stroke panel. *(Stroke weight .5 pt)*

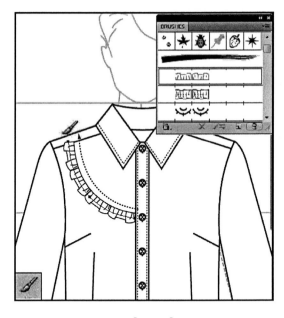

create pattern brush

1. Click to show the **Brushes** panel.

2. **Drag and drop** the ruffle tile into the **Brushes** panel.

3. In the **New Brush** dialog box click the **New Pattern Brush** button.

4. In the **Pattern Brush Options** dialog box type a name for the new pattern brush.

5. Click **OK**.

create brush stroke

1. Drag along the page to create a freehand brush stroke with the **Paintbrush** tool.

OR

1. Select a path then click the ruffle pattern brush in the **Brushes** panel.

explore details and trim

Detail or trim whether a ruffle, ribbing, tie string or even an entirely unique detail often times is the driving force behind an interesting design group. Create a new detail and try it out on your fashion flat, oh and beware...the possibilities could be endless.

chapter ten

draw a tailored jacket
advanced level lesson

By now you should be familiar with most of the drawing techniques that will make creating the tailored jacket an easy task.

In this chapter you will create a button front, long sleeve, tailored jacket with a lapel and welt pockets. You'll approach the tailored jacket by creating closed shapes for the right and left jacket that will overlap at the center front. This technique is used to create various versions of the tailored jacket, coat styles and vest styles. Of course you will use the front view of the tailored jacket to create its back view and then you will group the completed front and back views. To complete the tailored jacket you will add a herringbone pattern, located in the Swatches panel, and also rotate the pattern to match the sleeve grain.

create jacket shape

Just a reminder: Open a new Woman Croqui document, setup your workspace and remember to click the Default Fill and Stroke option before creating the jacket shapes. The tailored jacket is made up of three basic shapes: the left jacket shape, the right jacket shape and the back lining which is sent to back. The three shapes overlap leaving room for the buttons to be placed at the CF.

The jacket is made with three closed shapes.

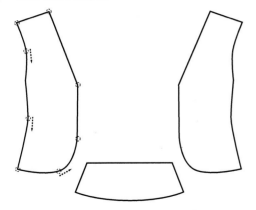

The shapes have a white fill and are stacked.

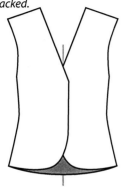

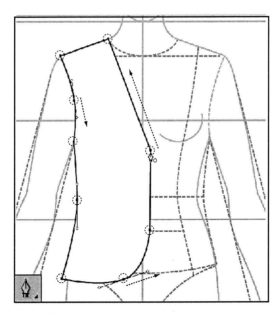

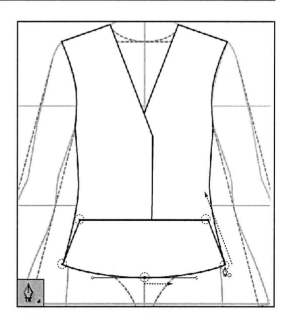

create jacket shape

1. Click **Default Fill and Stroke** in the **Tools** panel.

2. Draw the jacket shape with the **Pen** tool.

3. Smooth the jacket shape with the **Convert Anchor Point** tool.

4. Reshape the jacket shape with the **Direct Selection** tool.

reflect

1. Select the jacket shape with the **Selection** tool.

2. **Alt+click** at the CF with the **Reflect** tool.

3. Choose **Object > Arrange > Send to Back**.

4. Deselect with the **Selection** tool.

create lining shape

1. Draw the lining shape with the **Pen** tool.

2. Smooth the lining shape with the **Convert Anchor Point** tool.

3. Reshape the lining shape with the **Direct Selection** tool.

send to back

1. Select the back jacket shape with the **Selection** tool.

2. Choose **Object > Arrange > Send to Back**.

3. Deselect with the **Selection** tool.

add style lines

The Line Segment tool is an ideal drawing tool to use to create styles lines because it automatically turns off the fill. As a reminder, style lines and topstitch should not have a fill. To create the princess seam you will draw a straight line with the Line Segment tool then use the Convert Anchor Point tool to convert it into a smooth princess seam.

Style lines created with the Line Segment and Convert Anchor Point.

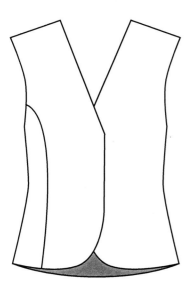

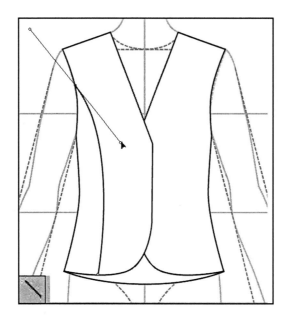

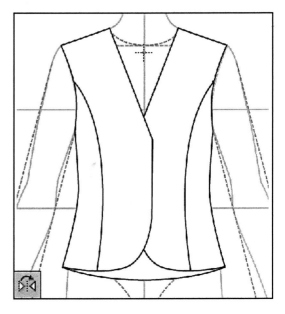

create style lines

1. Draw a style line with the **Line Segment** tool.

2. Smooth the style line with the **Convert Anchor Point** tool.

3. Reshape the style line with the **Direct Selection** tool.

reflect

1. Select the style line with the **Selection** tool.

2. **Alt+click** at the CF with the **Reflect** tool.

3. Deselect with the **Selection** tool.

create welt pocket

To create the welt pocket you will use the Rectangle tool and the Line segment tool remembering to create the detail with a white fill in order to stack it on top of the princess line.

There are many pocket options for a jacket of this style. You can use the various shape tools to create a patch pocket or even a flap pocket with topstitch detail. The Knife tool can be used to slice a rounded rectangle in half to create a flap pocket.

Slice across a rounded rectangle to create a flap pocket. Hold the Shift and Alt keys to carve a straight slice.

✗ to make more square

use down arrow — more round = up arrow

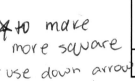

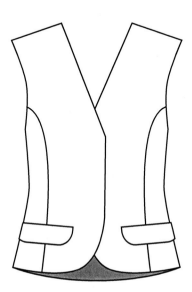

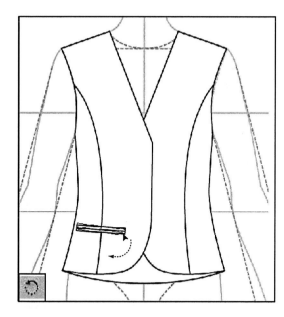

create welt pocket

1. Click **Default Fill and Stroke** in the **Tools** panel.

2. Draw the welt pocket shape with the **Rectangle** tool.

3. **Shift+drag** with the **Line Segment** tool to create a straight line across the rectangle.

group and scale

1. Select the complete pocket with the **Selection** tool.

2. Choose **Object > Group**.

3. **Shift+drag** the corner handle to scale the pocket.

rotate

1. Select the pocket with the **Selection** tool.

2. Select the **Rotate** tool.

3. Drag the pocket to rotate it.

reflect

1. **Alt+click** at the CF with the **Reflect** tool.

2. Deselect with the **Selection** tool.

create lapel

Similar to the shirt collar that was created in chapter six, the lapel is also made up of four basic shapes. The left lapel will be stacked on top of the right lapel shape, the back collar shape, and the lining shape. The Eraser tool is used to erase the exposed tip of the right lapel. To complete the lapel use the Line Segment tool to draw in the lapel seams.

The lapel is made with four basic shapes.

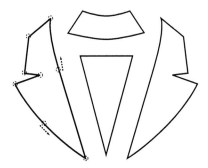

The shapes have a white fill and are stacked.

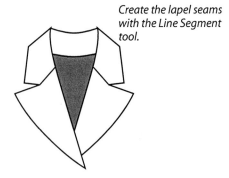

Create the lapel seams with the Line Segment tool.

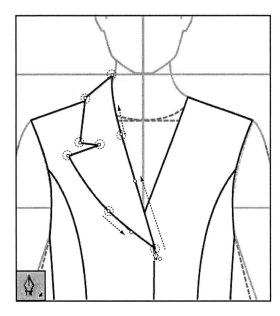

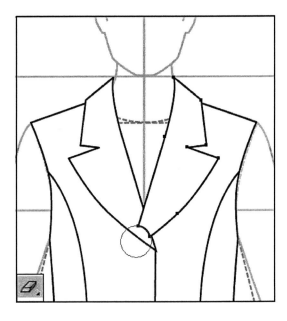

create lapel shape

1. Click the **Default Fill and Stroke** option in the **Tools** panel.

2. Draw the lapel shape with the **Pen** tool.

3. Smooth the lapel shape with the **Convert Anchor Point** tool.

4. Reshape the lapel shape with the **Direct Selection** tool.

reflect

1. Select the lapel shape with the **Selection** tool.

2. **Alt+click** at the CF with the **Reflect** tool.

3. Deselect with the **Selection** tool.

erase

1. Select the reflected lapel with the **Selection** tool.

2. Double-click the **Eraser** tool to set the eraser's angle, roundness and diameter. Click **OK**.

3. Drag the eraser accross the reflected lapel to erase the exposed tip.

bring to front

1. Select the original lapel shape with the **Selection** tool.

2. Choose **Object > Arrange > Bring to Front**.

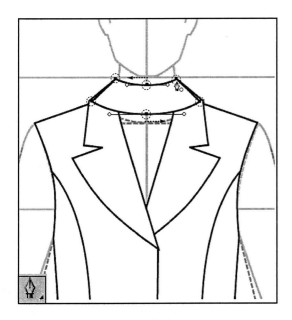

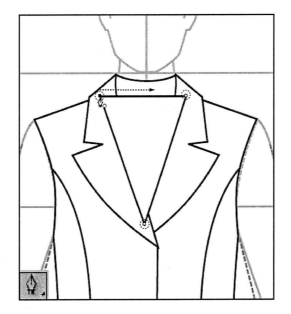

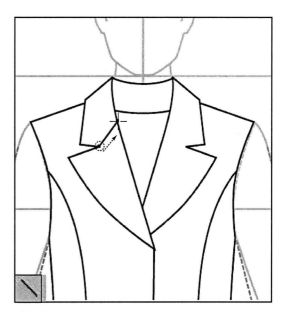

create back lapel shape

1. Draw the back lapel shape with the **Pen** tool.

2. Smooth the back lapel shape with the **Convert Anchor Point** tool.

send to back

1. Select the back lapel shape with the **Selection** tool.

2. Choose **Object > Arrange > Send to Back**.

create inside lapel shape

1. Draw the inside lapel shape with the **Pen** tool.

send to back

1. Select the inside lapel shape with the **Selection** tool.

2. Choose **Object > Arrange > Send to Back**.

create lapel seam

1. Draw the lapel seam with the **Line Segment** tool.

reflect

1. Select the seam with the **Selection** tool.

2. **Alt+click** at the CF with the **Reflect** tool.

3. Deselect with the **Selection** tool.

create button

If you have completed buttons in previous chapters they can be used here again for the jacket. Simply open a previously drawn fashion flat, ungroup it if needed then Copy/ Paste it unto this jacket. Buttons are created with simple shape tools, the Ellipse tool, the Rectangle tool and the Line Segment tool. To create a new button draw the button oversize, group and scale it to fit your jacket then set the stroke Weight to 0.5 pt in the Stroke panel.

The button is made with a few basic shapes.

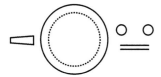

The shapes have a white fill and are stacked.

Group and scale the button then set the stroke Weight to 0.5 pt in the Stroke panel.

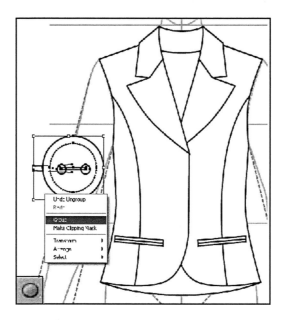

create button

1. Click the **Default Fill and Stroke** option in the Tools panel.

2. Click the **Ellipse** tool then hold the **Shift** and **Alt keys** on your keyboard.

3. Draw the button and its detail with the **Ellipse** tool and the **Line Segment** tool.

group and scale

1. Select the complete button with the **Selection** tool.

2. Choose **Object > Group**.

3. **Shift+drag** the corner handle to scale the button.

4. Set the stroke **Weight** in the **Stroke** panel.

duplicate button

1. Select the **Selection** tool then hold the **Shift key** and **Alt key** on your keyboard.

2. **Shift+Alt+drag** the button to duplicate it. *(Mac) Shift+Option+drag*

3. Press **Ctrl+D** on the keyboard to **Step Repeat**. Repeat Ctrl+D for the number of buttons desired. *(Mac) Cmd+D*

QUICK TIP!
Where you have several buttons that are grouped individually, select them then use the Align panel to Align Objects and Distribute Spacing.

---

create sleeve

Here you will create a long, set in sleeve. To create the shape of the sleeve you will copy the jacket's armhole path, paste it directly on top of the bodice armhole then use the Pen tool to pick up the path and close the shape. Sleeve detail such as a sleeve seam or topstich can be created with the Line Segment tool.

A basic sleeve shape is created with the Pen tool. A seam is added with the Line Segment tool.

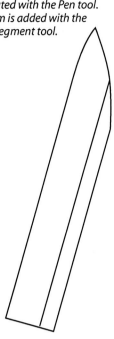

copy/ paste path

1. Select the **Direct Selection** tool.

2. Click to select the **middle armhole anchor**.

3. Choose **Edit > Copy** then **Edit > Paste In Front**.

create sleeve

1. Select the **Pen** tool.

2. Click inside the **shoulder armhole anchor** to pick up the path.

3. Draw the sleeve shape with the **Pen** tool.

4. Click inside the **armpit anchor** to close the sleeve shape.

create sleeve seam

1. Select the **Line Segment** tool.

2. Draw the sleeve seam with the **Line Segment** tool.

reflect

1. Select the complete sleeve with the **Selection** tool.

2. **Alt+click** at the CF with the **Reflect** tool.

3. Deselect with the **Selection** tool.

create back view

Now to create the back view of your tailored jacket. The back view of the tailored jacket is created by combining the left jacket shape, the right jacket shape and the lining shape. From there you will use various drawing tools to draw the back view style lines and detail.

The tailored jacket back view is created by combining the three jacket shapes.

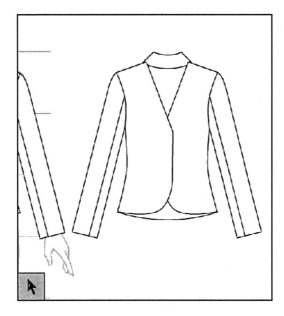

copy/ paste tailored jacket

1. Select the front view jacket with the **Selection** tool.

2. Choose **Edit > Copy** then **Edit > Paste**.

3. Move the copied jacket so that it is next to the original.

delete details

1. Select and delete the lapel shapes leaving only the back lapel shape.

2. Select and delete the pockets, buttons and style lines leaving only the jacket's right, left and lining shape.

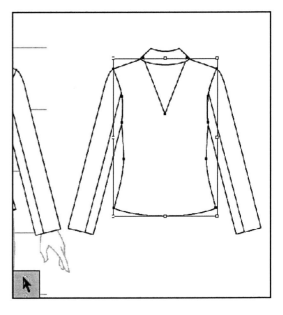

bring to front

1. Select the back collar with the **Selection** tool.

2. **Right-click** and choose **Arrange > Bring to Front.**

unite

1. Select the right and left jacket shapes and the lining shape with the **Selection** tool.

2. In the **Pathfinder** panel click the **unite** button.

reshape

1. Use the **Direct Selection tool** to move the **CB neckline anchor** (bodice) upward so that it is hidden beneath the back collar shape.

add back style lines

Adding detail to the back view of this tailored jacket is simple with the Line Segment tool. For the back sleeve borrow buttons from the front view jacket. There are many design options for the back view of a tailored jacket. Be certain to research and explore back views for optimal creativity.

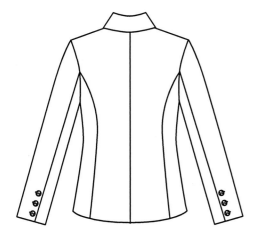

Add back view style lines with the Line Segment tool.

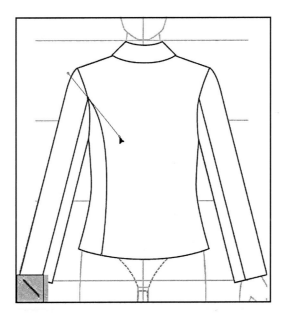

create back style lines

1. Draw the back view style line with the **Line Segment** tool.

2. Smooth the style line with the **Convert Anchor Point** tool.

3. Reshape the style line with the **Direct Selection** tool.

reflect

1. Select the style line with the **Selection** tool.

2. **Alt+click** at the CF with the **Reflect** tool.

3. Deselect with the **Selection** tool.

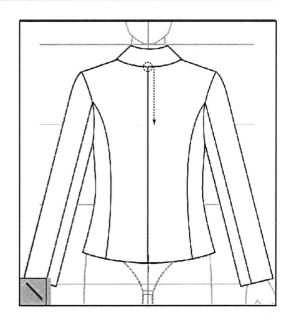

create back style lines

1. Draw the CB style line with the **Line Segment** tool.

   Note: Hold the **Shift** key while drawing to create a straight line.

group details/ group items

Now that you have completed your tailored jacket front and back views, you can group its seperate details. This will allow you to freely move fashion flats as one object, overlap them as in a presentation and easily ungroup them to explore new designs.

An ungrouped and exploded view of the tailored jacket with seperately grouped details.

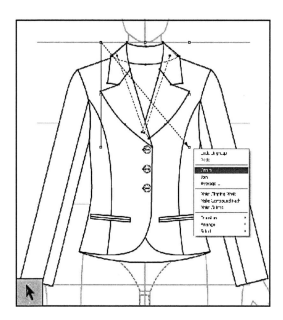

group front lapel

1. Select the **Selection** tool.

2. Drag a selection marquee from outside the jacket then down and over to select the lapel.

3. **Shift+click** the unwanted jacket parts to release them from the selection.

4. Choose **Object > Group**.

QUICK TIP!
Select several objects then Right-click on the artboard to choose the Group command. *(Mac) Control-click*

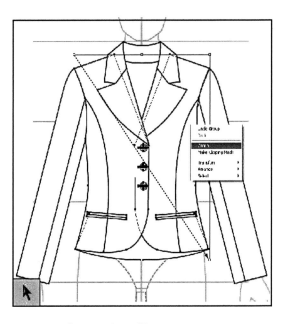

group front bodice

1. Drag a selection marquee from outside the jacket then down and over to select the front bodice.

2. **Shift+click** the lapel and/or the sleeve to release them from the selection.

3. Choose **Object > Group**.

group front sleeves

1. Drag a selection marquee to select the sleeve and its style lines.

2. Choose **Object > Group**.

   Note: Group the right sleeve then the left sleeve.

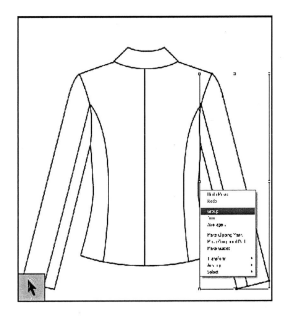

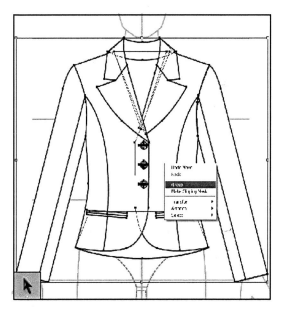

group back bodice

1. Drag a selection marquee from outside the jacket then down and over to select the back bodice.

2. **Shift+click** the lapel and/or the sleeve to release them from the selection.

3. Choose **Object > Group**.

group back sleeves

1. Drag a selection marquee to select the sleeve and its style lines.

2. Choose **Object > Group**.

   Note: Group the right sleeve then the left sleeve.

group entire front

1. Drag a selection marquee to select the jacket front view

2. Choose **Object > Group**.

group entire back

1. Drag a selection marquee to select the jacket back view

2. Choose **Object > Group**.

add pattern

The woman croqui's Swatches panel contains a herringbone pattern that will be used with this tailored jacket. To add the herringbone pattern to the tailored jacket simply drag it from the Swatches panel over to the closed portions of the jacket (bodice, sleeves, lapel).

rotate pattern

Once you've added the herringbone pattern you'll see that the grain of the pattern does not match the grain of the sleeve. Using the Tilde key and the Rotate tool you will rotate the pattern accordingly. The Tilde key (~) is located just below the Escape key on your keyboard. It is used to transform a pattern independant of the shape that it is in.

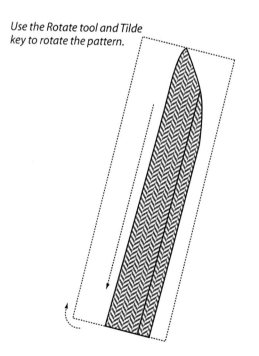

Use the Rotate tool and Tilde key to rotate the pattern.

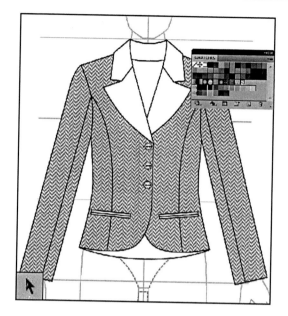

drag and drop pattern

1. Deselect with the **Selection** tool.

2. Click the herringbone pattern in the **Swatches** panel.

3. **Drag and drop** the pattern into the jacket shapes (bodice, sleeves).

QUICK TIP!
The Eyedropper tool allows you to sample stroke and fill attributes. Select an object then eyedrop the object you want to sample.

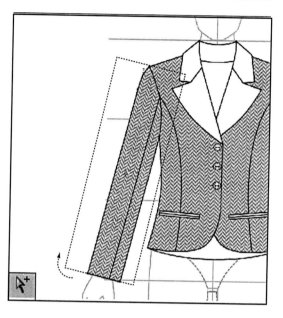

rotate pattern

1. Select a sleeve with the **Group Selection** tool.

2. Click the **Rotate** tool in the Tools panel.

3. Hold the **Tilde** key then drag the sleeve pattern to rotate it.

4. Select the opposite sleeve with the **Group Selection** tool. Use the Rotate tool and the Tilde key to rotate the pattern.

   Note: Rotate the right sleeve pattern then the left sleeve pattern.

explore tailored jacket designs

The techniques demonstrated with this tailored
jacket, for example drawing closed bodice
shapes that overlap, can be used to create the
various jacket styles, coat styles and vests styles.

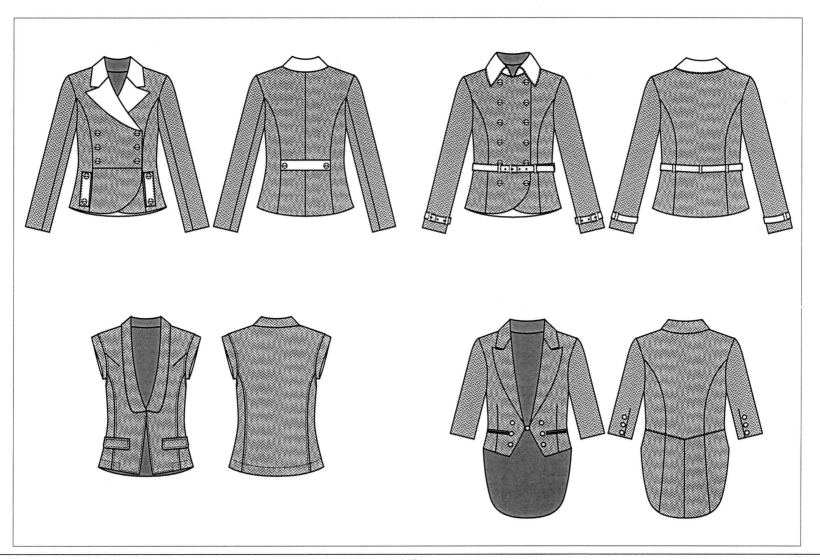

chapter eleven

and now presenting

A presentation of your fashion flats is an effective way of communicating themed ideas, demonstrating organized concepts and displaying fabric and color options. Buyers will review the presentation to get familiar with the products being developed by a manufacturer's design team. A clear and organized presentation is effective in letting merchandise buyers know what the designer's creative projections are.

In this chapter you will learn to use the various type tools and you'll explore creating stylized type for titles and logos. You will learn how to use shape tools to add simple backgounds and borders and you'll use the Swatches panel to create pattern and fabric swatches.

buttoned shirt with ruffled sleeves and patch pockets.

have a plan

Plan the presentation and decide what information will be included. A focused presentation may include a themed title and season, page graphics such as backgrounds or borders, garment descriptions, fabric descriptions, fabric swatches and colorways.

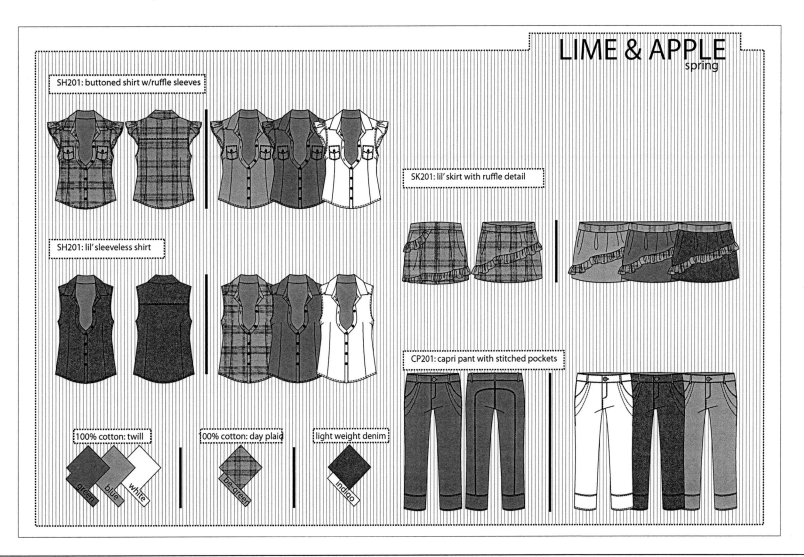

create type

The Type tool creates point type which stands alone, without a shape to control the area where type flows. With the Type tool you will press the Enter key on your keyboard to return the type. To create point type use the Type tool to click on a blank area of the page then type the desired information, title or logo.

The Area Type tool creates type within a shape. This tool is ideal for large areas of type as in paragraphs. To create area type use the Area Type tool to click on a shape then type the desired information. Use the Selection tool to select then edit type, changing its color, font family, font style and/or size in the Control panel. The Character panel offers more options for editing type such as leading and kerning.

The Control panel offers easy access to type attributes.

Click to open the Character panel.

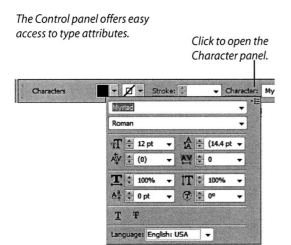

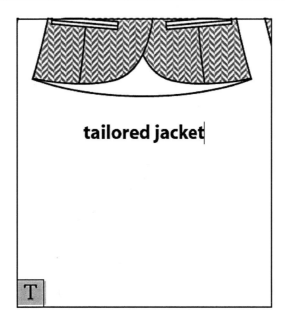

create type

1. Select the **Type** tool.

2. Click on a blank area of the page then type the desired information.

edit type

1. Select the **Selection** tool.

2. Click on the edge of the type to select it.

   OR

2. Select the **Type** tool then drag (highlight) an area of type.

3. Click an option in the **Control Panel** to edit the selected type.

create area type

1. Create a basic shape (**Rectangle or Ellipse**).

2. Select the **Area Type** tool.

3. Click on a path of the shape then type the desired information.

> **QUICK TIP!**
> The Type tool automatically sets the fill to black and the stroke to none. Remember to set the stroke/fill option to its default attributes before creating a shape or fashion flat.

create stylized type

Adobe Illustrator has a thorough selection of type and editing tools that make designing a logo interesting and fun too. The thing to remember about creating stylized type is that it be legible at various sizes and that it clearly convey the fashion message. Explore logo design, try out various fonts, stroke/ fill colors, patterns, stroke weights and warp styles then choose your favorite to represent your fashion style.

Stylized type created with type tools and the Make with Warp command.

romantic fashion

Font: Fiolex Girls/ Warp style: None

Classic Fashion

Font: Ariel Narrow/ Warp style: None

Font: Ravie/ Warp style: Bulge

Font: Rope/ Warp style: Shell Uppper

fantastic fashion and style

Font: Quigley Wiggly/ Type tool: Type on a Path

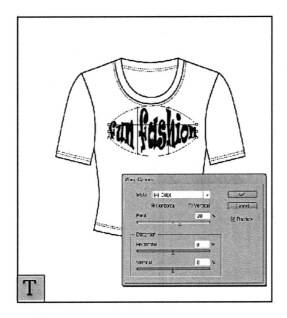

create type

1. Create point type (see page 145).

edit type

1. Select the **Selection** tool.

2. Click on the edge of the type to select it.

3. Click an option in the **Control Panel** to edit the selected type.

make with warp

1. Select the type with the **Selection** tool.

2. Choose **Object > Envelope Distort > Make with Warp**. Choose a warp Style.

3. Click **OK**.

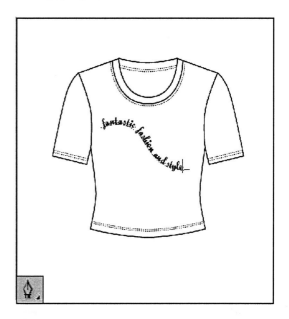

type on a path

1. Draw a smooth path with the **Pen** tool.

2. Select the **Type on a Path** tool from the **Tools** panel.

3. Click on the path then type the desired information.

edit type

1. Select the **Type** tool then drag (highlight) an area of type.

2. Click an option in the **Control Panel** to edit the selected type.

add background

Creating background colors or patterns to compliment your fashion layout is simple using any of the shape tools. Ideally when adding a background to your layout you should create a new layer to keep the background contents seperate from your fashion flats. Once the background shape is created on the new layer you will then apply color, gradients or patterns that relate to your fashion theme and layout.

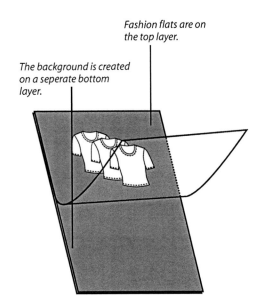

Fashion flats are on the top layer.

The background is created on a seperate bottom layer.

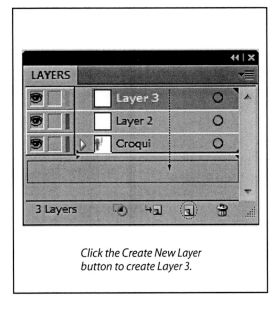

Click the Create New Layer button to create Layer 3.

create new layer

1. Click to show the **Layers** panels.

2. Click the **Create New Layer** button.

3. Drag the new layer to the bottom of the layer stack.

4. Double click the new layer and title it "**background layer**".

QUICK TIP!
The Transparency panel allows you to blend shapes and colors by turning down the opacity of the object. Explore the Transparency panel by creating two or more objects and then changing their Opacity.

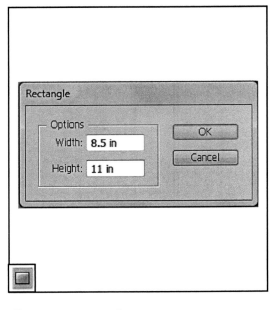

draw rectangle

1. Select the **Rectangle** tool.

2. Click on the page to open the **Rectangle** tool dialog box.

3. Type in a measurement that matches the size of the page. (8.5 in x 11 in)

4. Click **OK** then use the **Selection** tool to position the rectangle on the page.

Note: Once you have completed the background and its color you can then lock the background layer in the Layers panel.

create border

The Brushes panel offers a selection of border styles that can be applied to any shape. To create a page border for your fashion layout simply create a rectangle then apply a border brush from the Brushes panel. You will find additional border brushes in the Brushes Library. To open the Brushes Library click the Brush panel menu button then choose Open Brush Library. Remember that most printers will not print to the edge of the page. For this reason you will need to create a rectangle that fits just inside the Imageable Area. Be certain the Page Tiling is showing in the View menu.

The Brushes panel offers a selection of border styles.

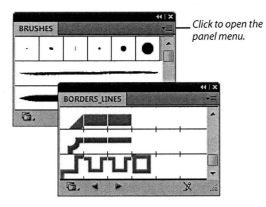

Click to open the panel menu.

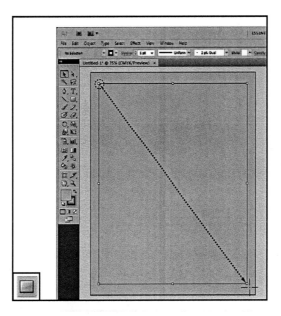

draw rectangle

Note: You can create this rectangle on the new and seperate layer.

1. Select the **Rectangle** tool.

2. Draw a rectangle with the **Rectangle** tool.

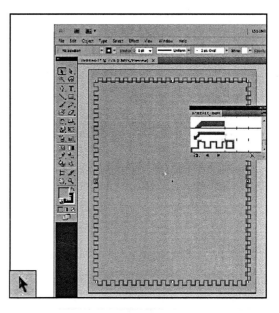

apply border brush

1. Select the rectangle with the **Selection** tool.

2. Click to show the **Brushes** panel.

3. Click the Brushes panel menu button.

4. Choose **Open Brush Library > Borders > Borders_Dashed**.

QUICK TIP!
To increase or decrease the size of a border brush change its stroke Weight in the Stroke panel.

---

swatches panel

The Swatches panel stores colors, gradients and patterns that are used to colorize and fabricate fashion flats. To create a new pattern swatch you will first draw a background color for the pattern with the Rectangle tool then use various drawing tools to create the pattern tile. Ideally you will want to draw the pattern tile oversize then scale it down to the actual size that you want it to appear as in your fashion flat.

Create one pattern tile (repeating detail).

Pattern swatches created with the Swatches panel.

define pattern

Once you have created the pattern tile you will want to define the pattern so that it will repeat without a gap between each tile. To do this you will draw a rectangle that fits just inside the background color. Turn off the rectangle's fill and stroke so as to create an invisible rectangle and then send it to the back of the tile (Object > Arrange > Send to Back).

Define the pattern with an invisible rectangle that is sent to back.

A flounce skirt with a polka dot pattern.

> **QUICK TIP!**
> Once the pattern is applied to the fashion flat use the Tilde key on your keyboard to transform the pattern. Use the Tilde key and the Scale tool to scale the pattern. Use the Tilde key and the Rotate tool to rotate the pattern.

create pattern swatch

Once you have created the pattern tile and defined it with an invisible rectangle, drag all its parts into the Swatches panel to create a New Pattern Swatch. To edit the pattern swatch drag the pattern from the Swatches panel and drop it onto a blank area on the page, change the desired colors, scale its size and/or add more detail. Drag the new pattern tile back into the Swatches panel to create a brand new pattern swatch.

Scale the tile then drag it into the Swatches panel to create a New Pattern Swatch.

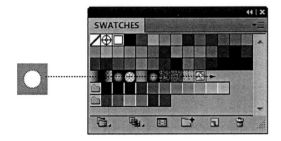

To edit a pattern swatch drag it from the swatches panel onto a blank area on the page.

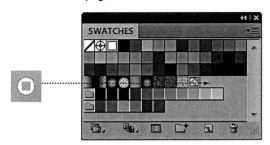

Drag the new pattern tile back into the Swatches panel.

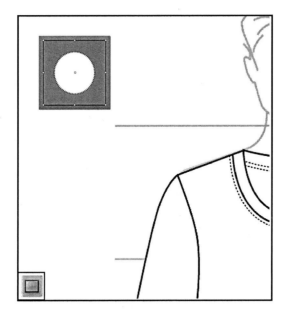

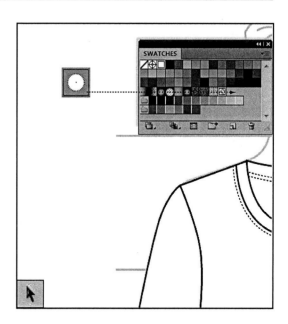

create pattern tile

1. Draw a background color for the pattern with the **Rectangle** tool.

2. Draw the desired pattern with various drawing and shape tools.

define pattern tile

1. Draw a rectangle on top of the pattern with the **Rectangle** tool.

2. Set the **Fill** color to **None** in the **Control** panel.

3. Set the **Stroke** color to **None** in the **Control** panel.

4. Choose **Object > Arrange > Send to Back**.

scale pattern tile

1. Select the entire pattern tile with the **Selection** tool then hold the **Shift** key.

2. **Shift+drag** the corner handle to scale the pattern tile to the desired size.

create pattern swatch

1. Click to show the **Swatches** panel.

2. **Drag and drop** the pattern tile into the **Swatches** panel.

scan fabric

To scan a fabric into the computer you will need a scanner, its software and Adobe Photoshop. This tutorial will demonstrate how to use Adobe Photoshop to scan a fabric. Not all computers are setup the same way so for more information on scanning fabric or images refer to your scanner's users manual.

Adobe Photoshop has its own set of painting tools, editing tools and image correction menus. For more detailed information on using Photoshop to paint, edit and touchup images refer to your Adobe Photoshop user manual.

Keep in mind that scanning fabric into a paint program such as Photoshop is much like making a digital photocopy of the fabric. Here you will scan a fabric, save it then place it into Adobe Illustrator so that it can be made into a fabric swatch.

Position the fabric as straight and flat as possible and face down on the scanner bed.

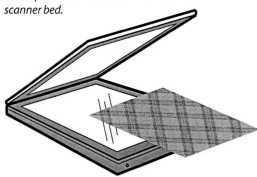

Choose File > Import > (WIA-your scanner) to open the Scan dialog box.

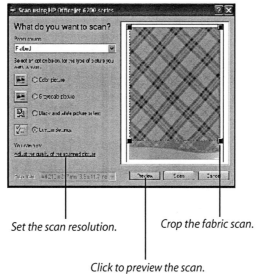

Set the scan resolution.　　*Crop the fabric scan.*

Click to preview the scan.

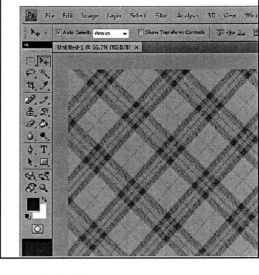

scan fabric

1. Position the fabric on the scanner bed.

2. Launch **Adobe Photoshop**.

3. Choose **File > Import** then select the scanner that is connected to your computer.

   Note: The type of scanner and its software will vary. In general you will scan a color document at a resolution of 150 dpi (dots per inch).

4. Click **Preview**.

5. Crop the fabric within the scan window.

6. Click **Scan**.

7. In Photoshop choose **File > Save**.

8. In Photoshop choose **File > Exit**.

create fabric swatch

Once the fabric is scanned and saved as a Photoshop file you will then use Adobe Illustrator's File menu to Place the fabric. Be certain to turn off the Link option in the Place dialog box. Turning off the link option will bring the fabric into the Illustrator document without a link to the original document.

The fabric is scanned as its actual size, much larger than most fashion flats, and so you will need to scale the fabric down to the actual size that you want it to appear as in your fashion flat. To create the New Fabric Swatch drag the fabric into the Swatches panel.

Drag the fabric into the Swatches panel to create a New Fabric Swatch.

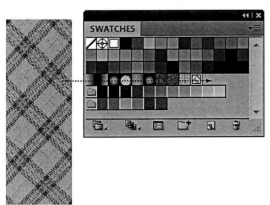

place fabric

1. In Illustrator choose **File > Place**.

2. In the **Place** dialog box navigate to the saved Photoshop document.

3. In the **Place** dialog box uncheck the **Link** option.

4. Click **Place.**

scale fabric

1. Select the placed fabric with the **Selection** tool then hold the **Shift** key.

2. **Shift+drag** the corner handle to scale the fabric to the desired size.

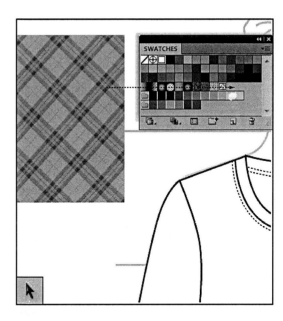

create fabric swatch

1. Click to show the **Swatches** panel.

2. **Drag and drop** the fabric into the **Swatches** panel.

QUICK TIP!
Once the fabric is applied to the fashion flat use the Tilde key on your keyboard to transform the fabric. Hold the Tilde key and use the Scale tool to scale the fabric. Hold the Tilde key and use the Rotate tool to rotate the fabric.

---

presentation samples

This trend presentation features men's sports-
wear items along with fabric swatches, colorways, page graphics and a themed title.

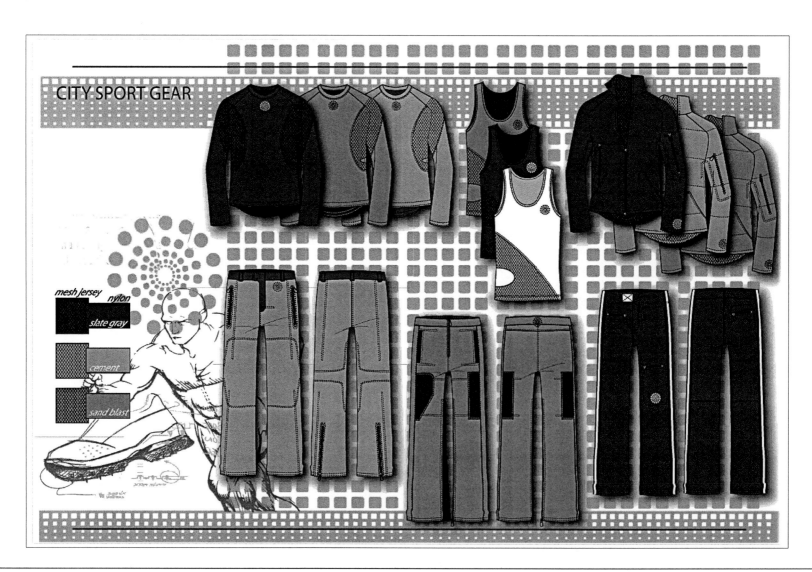

presentation samples

This itemized presentation includes women's casual items, style numbers, fabric information, size ranges and a themed title. The fabric swatches and colorways may accompany this type of presentation.

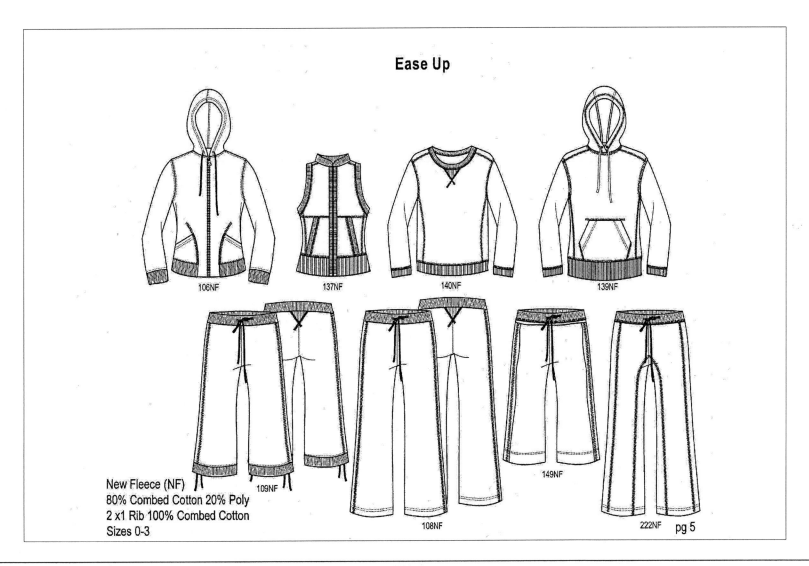

Ease Up

New Fleece (NF)
80% Combed Cotton 20% Poly
2 x1 Rib 100% Combed Cotton
Sizes 0-3

pg 5

fashion flat gallery

be inspired

Pants and Jackets and Skirts..........oh my! There is so much to be inspired by from runway fashion to street art and architecture too. Once you have a concept and theme for your design render a quick or working sketch, maybe even on that napkin at the diner. Plan your designs with fabric, color, themes and silhouettes in mind and you'll find that drawing the electric fashion flat happens with great efficiency, precision and endless possibilities.

details

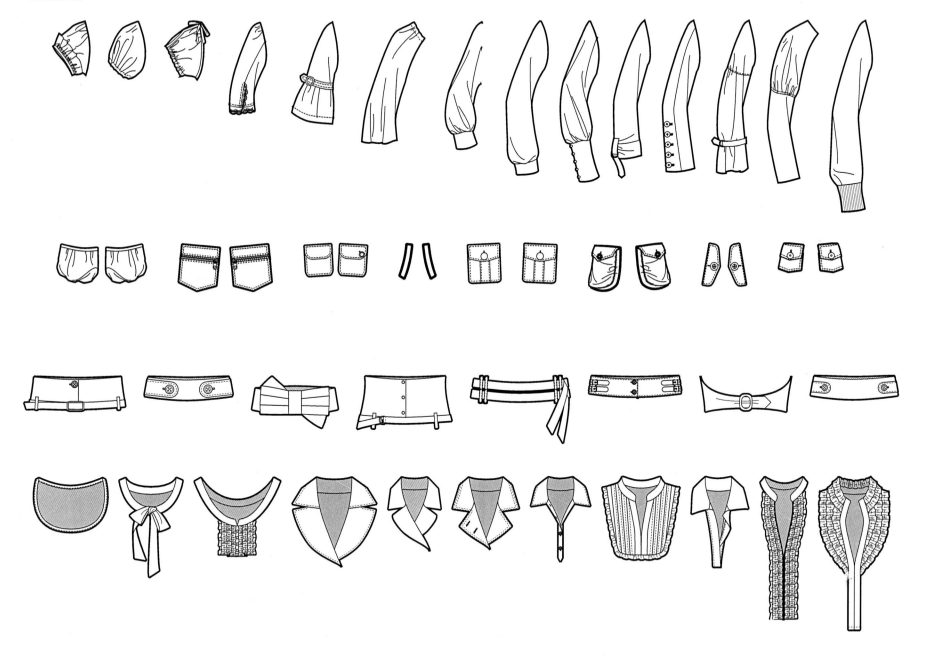

tops

shirts/ blouses

shorts

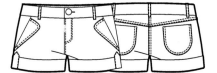
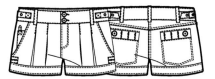
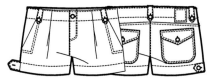
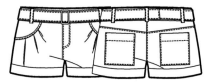

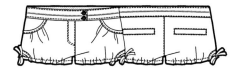
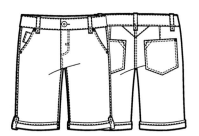
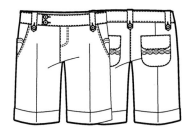
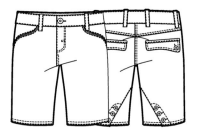

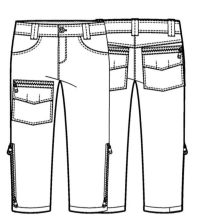

pants

jumpers

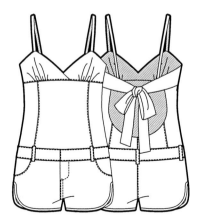
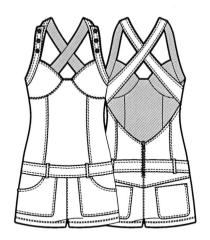

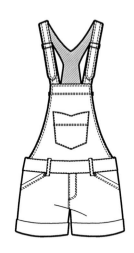

skirts

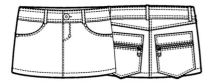

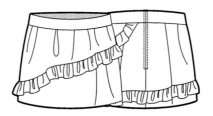
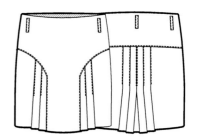

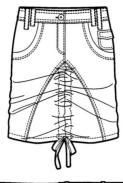
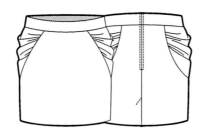

vests/ jackets

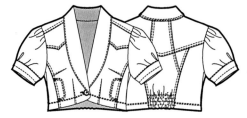

coats

dresses

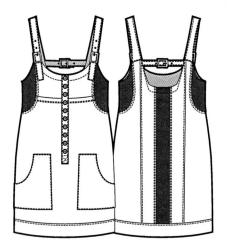

dresses

index